Bisphenol A
A Multi-modal Endocrine Disruptor

Issues in Toxicology

Editor-in-chief:
Diana Anderson, *University of Bradford, UK*

Series editors:
Michael D. Waters, *Michael Waters Consulting, USA*
Timothy C. Marrs, *Edentox Associates, UK*
Alok Dhawan, *Centre of Biomedical Research, India*

Titles in the Series:

1: Hair in Toxicology: An Important Bio-Monitor
2: Male-mediated Developmental Toxicity
3: Cytochrome P450: Role in the Metabolism and Toxicity of Drugs and other Xenobiotics
4: Bile Acids: Toxicology and Bioactivity
5: The Comet Assay in Toxicology
6: Silver in Healthcare
7: *In Silico* Toxicology: Principles and Applications
8: Environmental Cardiology
9: Biomarkers and Human Biomonitoring, Volume 1: Ongoing Programs and Exposures
10: Biomarkers and Human Biomonitoring, Volume 2: Selected Biomarkers of Current Interest
11: Hormone-Disruptive Chemical Contaminants in Food
12: Mammalian Toxicology of Insecticides
13: The Cellular Response to the Genotoxic Insult: The Question of Threshold for Genotoxic Carcinogens
14: Toxicological Effects of Veterinary Medicinal Products in Humans: Volume 1
15: Toxicological Effects of Veterinary Medicinal Products in Humans: Volume 2
16: Aging and Vulnerability to Environmental Chemicals: Age-related Disorders and their Origins in Environmental Exposures
17: Chemical Toxicity Prediction: Category Formation and Read-Across
18: The Carcinogenicity of Metals: Human Risk Through Occupational and Environmental Exposure
19: Reducing, Refining and Replacing the Use of Animals in Toxicity Testing
20: Advances in Dermatological Sciences
21: Metabolic Profiling: Disease and Xenobiotics
22: Manganese in Health and Disease
23: Toxicology, Survival and Health Hazards of Combustion Products
24: Masked Mycotoxins in Food: Formation, Occurrence and Toxicological Relevance
25: Aerobiology: The Toxicology of Airborne Pathogens and Toxins

26: Chemical Warfare Toxicology, Volume 1: Fundamental Aspects
27: Chemical Warfare Toxicology, Volume 2: Management of Poisoning
28: Toxicogenomics in Predictive Carcinogenicity
29: Human Stem Cell Toxicology
30: The Comet Assay in Toxicology, 2nd edition
31: Computational Systems Pharmacology and Toxicology
32: Ecotoxicology and Genotoxicology: Non-traditional Terrestrial Models
33: Ecotoxicology and Genotoxicology: Non-traditional Aquatic Models
34: Oxidative Stress and Redox Signalling in Parkinson's Disease
35: Nanotoxicology: Experimental and Computational Perspectives
36: Regulatory Toxicology in the European Union
37: Formaldehyde: Exposure, Toxicity and Health Effects
38: Chemical Health Threats: Assessing and Alerting
39: The Micronucleus Assay in Toxicology
40: Nanoparticle–Protein Corona: Biophysics to Biology
41: Big Data in Predictive Toxicology
42: Challenges in Endocrine Disruptor Toxicology and Risk Assessment
43: Bisphenol A: A Multi-modal Endocrine Disruptor

How to obtain future titles on publication:
A standing order plan is available for this series. A standing order will bring delivery of each new volume immediately on publication.

For further information please contact:
Book Sales Department, Royal Society of Chemistry, Thomas Graham House, Science Park, Milton Road, Cambridge, CB4 0WF, UK
Telephone: +44 (0)1223 420066, Fax: +44 (0)1223 420247
Email: booksales@rsc.org
Visit our website at www.rsc.org/books

Bisphenol A
A Multi-modal Endocrine Disruptor

Edited by

Natalie R. Gassman
University of Alabama at Birmingham, USA
Email: nrg2@uab.edu

ROYAL SOCIETY
OF **CHEMISTRY**

Issues in Toxicology No. 43

Print ISBN: 978-1-83916-206-0
PDF ISBN: 978-1-83916-649-5
EPUB ISBN: 978-1-83916-650-1
Print ISSN: 1757-7179
Electronic ISSN: 1757-7187

A catalogue record for this book is available from the British Library

The Royal Society of Chemistry is a charity, registered in England and Wales, Number 207890, and a company incorporated in England by Royal Charter (Registered No. RC000524), registered office: Burlington House, Piccadilly, London W1J 0BA, UK, Telephone: +44 (0) 20 7437 8656.

For further information see our web site at www.rsc.org

Printed in the United Kingdom by CPI Group (UK) Ltd, Croydon, CR0 4YY, UK

Preface

The discovery of plasticizers and their myriad uses has revolutionized our society. These compounds have created an age of convenient single-use products, improved medical devices, and advanced consumer products. We come into contact with plastics and plasticizers on almost a daily basis, and as a result, they are some of the most highly produced chemical entities in the world. Bisphenol A (BPA) is one such plasticizer, whose utility and chemical properties have made it highly amenable to consumer product use. As a result, millions of tons of BPA are produced each year, which has led to widespread human exposure through the air, water, soil, and consumer products.

The endocrine-disrupting properties of BPA were discovered early on, given its phenolic nature and similarity to steroid structures. However, its use since its discovery in the 1940s has only continued to grow. Concern about BPA and other plasticizers reached the mainstream media in the early 2000s, prompting regulatory reviews of its use in consumer products and, in particular, its use in infant products. In 2007, US regulatory agencies evaluated BPA toxicology data. They determined that nanomolar levels of BPA exposure was safe for adults but noted "some concern" about the effects of BPA on fetuses, infants, and children.[1] Since this decision, hundreds of scientific studies have been published that continue to question the safety of BPA. Critically, there is a significant weight of evidence that BPA exposures in the 10–100 nM range have adverse outcomes, which include inducing reproductive toxicity, altering metabolism, affecting neurobehaviors, and sensitizing tissues to carcinogenic challenges.[2–7] The carcinogenic potential of BPA has been a growing focus area with epidemiological studies linking BPA exposures with breast and liver cancer development[8–12] and neoplastic changes in other reproductive organs and tissues.[5] Despite this weight of evidence, the health risks associated with the chronic nanomolar BPA

Issues in Toxicology No. 43
Bisphenol A: A Multi-modal Endocrine Disruptor
Edited by Natalie R. Gassman

Published by the Royal Society of Chemistry, www.rsc.org

exposure experienced by the population are still controversial due to a lack of molecular mechanisms describing the pleiotropic effects of observed exposures.[13–17] The pleiotropic effects induced by BPA have made understanding its modes of action and characterizing the consequences of exposure challenging.

I first began studying BPA in 2012 under the guidance of Dr Samuel H. Wilson at the National Institute of Environmental Health Sciences. We were interested in the induction of DNA damage by BPA and its impact on DNA repair mechanisms. Our studies found that BPA induces DNA damage directly and indirectly through reactive oxygen species.[18,19] More importantly, we also found that BPA exposure alters DNA damage response and repair mechanisms, promoting the persistence of mutagenic DNA lesions within the genome.[18,19] These findings were especially interesting, because they occurred in cell-line models that lacked or have extremely low expression of estrogen receptors. Thus, we identified a critical role for the metabolism of BPA and its generation of reactive species in promoting adverse cellular consequences.[20]

Our work and that of numerous others have begun to identify modes of action for BPA that are not linked to the classical estrogen receptors. BPA's modes of action have been found to use other steroid receptors, non-classical estrogen receptors, and even direct interactions with signaling molecules like RAS.[21] It is becoming increasingly appreciated that BPA has multi-modal effects. In addition to non-monotonic dose responses, these effects have likely confounded our ability to understand the cellular and organismal consequences of BPA exposure.

While there are significant efforts across the globe to better understand our exposure to BPA and other endocrine-disrupting chemicals, the health risks associated with the chronic BPA exposure experienced by the human population are still unknown. The lack of molecular mechanisms describing the pleiotropic effects observed with BPA exposures significantly contributes to this problem and needs to be addressed by funding and regulatory agencies.[13–17] When I embarked on my first study of BPA, I did not understand the impact it would have on my research and worldview. BPA is a persistent environmental toxicant from which we will continue to experience exposure effects for decades to come. Its reproductive toxicity means our children and grandchildren will experience these effects as well. There is a wide global effort underway to understand these effects. In this book, we highlight the multi-modal mechanism of BPA and examine its cell and organismal consequences. As noted in most chapters in this volume, much more work is needed to understand better the impact of BPA on the genome, epigenome, and proteome. However, as the work highlighted in this volume shows, we have made progress in understanding BPA's multi-modal effects. The global research efforts highlighted here also demonstrate that we will continue to progress, and hopefully, eventually alleviate the impact that BPA and other endocrine-disrupting chemicals have on our population.

Natalie R. Gassman

References

1. R. E. Chapin, J. Adams, K. Boekelheide, L. E. Gray, S. W. Hayward, P. S. Lees, B. S. McIntyre, K. M. Portier, T. M. Schnorr, S. G. Selevan, J. G. Vandenbergh and S. R. Woskie, NTP-CERHR expert panel report on the reproductive and developmental toxicity of bisphenol A, *Birth Defects Res., Part B*, 2008, **83**, 157–395.
2. L. N. Vandenberg, S. Ehrlich, S. M. Belcher, N. Ben-Jonathan, D. C. Dolinoy, E. R. Hugo, P. A. Hunt, R. R. Newbold, B. S. Rubin, K. S. Saili, A. M. Soto, H.-S. Wang and F. S. vom Saal, Low dose effects of bisphenol A, *Endocr. Disruptors*, 2013, **1**, e26490.
3. J. R. Rochester, Bisphenol A and human health: a review of the literature, *Reprod. Toxicol.*, 2013, **42**, 132–155.
4. A. C. Gore, V. A. Chappell, S. E. Fenton, J. A. Flaws, A. Nadal, G. S. Prins, J. Toppari and R. T. Zoeller, Executive Summary to EDC-2: The Endocrine Society's Second Scientific Statement on Endocrine-Disrupting Chemicals, *Endocr. Rev.*, 2015, **36**, 593–602.
5. D. D. Seachrist, K. W. Bonk, S. M. Ho, G. S. Prins, A. M. Soto and R. A. Keri, A review of the carcinogenic potential of bisphenol A, *Reprod. Toxicol.*, 2016, **59**, 167–182.
6. O. S. Anderson, M. S. Nahar, C. Faulk, T. R. Jones, C. Liao, K. Kannan, C. Weinhouse, L. S. Rozek and D. C. Dolinoy, Epigenetic responses following maternal dietary exposure to physiologically relevant levels of bisphenol A, *Environ. Mol. Mutagen.*, 2012, **53**, 334–342.
7. J. H. Kim, L. S. Rozek, A. S. Soliman, M. A. Sartor, A. Hablas, I. A. Seifeldin, J. A. Colacino, C. Weinhouse, M. S. Nahar and D. C. Dolinoy, Bisphenol A-associated epigenomic changes in prepubescent girls: a cross-sectional study in Gharbiah, Egypt, *Environ. Health*, 2013, **12**, 33.
8. C. Weinhouse, O. S. Anderson, I. L. Bergin, D. J. Vandenbergh, J. P. Gyekis, M. A. Dingman, J. Yang and D. C. Dolinoy, Dose-dependent incidence of hepatic tumors in adult mice following perinatal exposure to bisphenol A, *Environ. Health Perspect.*, 2014, **122**, 485–491.
9. N. Acevedo, B. Davis, C. M. Schaeberle, C. Sonnenschein and A. M. Soto, Perinatally administered bisphenol a as a potential mammary gland carcinogen in rats, *Environ. Health Perspect.*, 2013, **121**, 1040–1046.
10. A. M. Betancourt, I. A. Eltoum, R. A. Desmond, J. Russo and C. A. Lamartiniere, In utero exposure to bisphenol A shifts the window of susceptibility for mammary carcinogenesis in the rat, *Environ. Health Perspect.*, 2010, **118**, 1614–1619.
11. M. Durando, L. Kass, J. Piva, C. Sonnenschein, A. M. Soto, E. H. Luque and M. Munoz-de-Toro, Prenatal bisphenol A exposure induces preneoplastic lesions in the mammary gland in Wistar rats, *Environ. Health Perspect.*, 2007, **115**, 80–86.
12. S. Jenkins, N. Raghuraman, I. Eltoum, M. Carpenter, J. Russo and C. A. Lamartiniere, Oral exposure to bisphenol a increases

dimethylbenzanthracene-induced mammary cancer in rats, *Environ. Health Perspect.*, 2009, **117**, 910–915.
13. L. Trasande, L. N. Vandenberg, J. P. Bourguignon, J. P. Myers, R. Slama, F. Vom Saal and R. T. Zoeller, Peer-reviewed and unbiased research, rather than 'sound science', should be used to evaluate endocrine-disrupting chemicals, *J. Epidemiol. Community Health*, 2016, **70**, 1051–1056.
14. L. N. Vandenberg and G. S. Prins, Clarity in the face of confusion: new studies tip the scales on bisphenol A (BPA), *Andrology*, 2016, **4**, 561–564.
15. P. Kovacic, How safe is bisphenol A? Fundamentals of toxicity: metabolism, electron transfer and oxidative stress, *Med. Hypotheses*, 2010, **75**, 1–4.
16. T. T. Schug, J. J. Heindel, L. Camacho, K. B. Delclos, P. Howard, A. F. Johnson, J. Aungst, D. Keefe, R. Newbold, N. J. Walker, R. Thomas Zoeller and J. R. Bucher, A new approach to synergize academic and guideline-compliant research: the CLARITY-BPA research program, *Reprod. Toxicol.*, 2013, **40**, 35–40.
17. L. N. Vandenberg and G. S. Prins, Clarity in the face of confusion: new studies tip the scales on bisphenol A (BPA), *Andrology*, 2016, **4**, 561–564.
18. N. R. Gassman, E. Coskun, P. Jaruga, M. Dizdaroglu and S. H. Wilson, Combined Effects of High-Dose Bisphenol A and Oxidizing Agent (KBrO3) on Cellular Microenvironment, Gene Expression, and Chromatin Structure of Ku70-deficient Mouse Embryonic Fibroblasts, *Environ. Health Perspect.*, 2016, **124**, 1241–1252.
19. N. R. Gassman, E. Coskun, D. F. Stefanick, J. K. Horton, P. Jaruga, M. Dizdaroglu and S. H. Wilson, Bisphenol A promotes cell survival following oxidative DNA damage in mouse fibroblasts, *PLoS One*, 2015, **10**, e0118819.
20. N. R. Gassman, Induction of oxidative stress by bisphenol A and its pleiotropic effects, *Environ. Mol. Mutagen.*, 2017, **58**, 60–71.
21. M. Schopel, K. F. Jockers, P. M. Duppe, J. Autzen, V. N. Potheraveedu, S. Ince, K. T. Yip, R. Heumann, C. Herrmann, J. Scherkenbeck and R. Stoll, Bisphenol A binds to Ras proteins and competes with guanine nucleotide exchange: implications for GTPase-selective antagonists, *J. Med. Chem.*, 2013, **56**, 9664–9672.

Contents

Issues in Toxicology No. 43
Bisphenol A: A Multi-modal Endocrine Disruptor
Edited by Natalie R. Gassman

Published by the Royal Society of Chemistry, www.rsc.org

CHAPTER 1

Classical and Non-classical Estrogen Receptor Effects of Bisphenol A

MANOJ SONAVANE

Department of Pharmacology and Toxicology, University of Alabama at Birmingham, Birmingham, AL, USA
Email: msonavane@uab.edu

1.1 Introduction

The word estrogen is commonly used to refer to 17β-estradiol (E2) due to its physiological relevance and predominance during reproductive growth. Estrogens are sex steroid hormones primarily synthesized in the ovaries and the adrenal glands, adipose tissue, brain, and testis. Estrogen displays a broad spectrum of physiological functions, including regulation of the menstrual cycle and reproduction, bone density, brain function, cholesterol mobilization, control of inflammation, and development of breast tissue and sexual organs.[1] While estrogens play diverse and similar physiological roles in both sexes,[2] they control primary and secondary sexual characteristics in females. In puberty, estradiol (E2) promotes epithelial cell proliferation and mammary glands, whereas estrogen helps prepare the mammary gland for milk production during pregnancy.[2–5] The lower levels of estrogens produced in men are essential for sperm maturation, erectile function, and healthy libido.[6] Besides this sexual and reproductive role, E2 exerts many actions in other systems such as the adipose tissue, bone, brain,

Issues in Toxicology No. 43
Bisphenol A: A Multi-modal Endocrine Disruptor
Edited by Natalie R. Gassman

Published by the Royal Society of Chemistry, www.rsc.org

cardiovascular system, endocrine system, pancreas, liver, and skeletal muscle.[7,8] It is important to note that any synthetic or semi-synthetic steroid that mimics the effects of natural estrogens is considered an estrogen.

1.2 Mechanism of Estrogen Signaling

All the physiological functions of estrogen are mediated *via* estrogen receptors (ERs). In 1958, Elwood Jensen discovered ERs by showing that female reproductive tissues could uptake estrogen from the circulation by binding to proteins. Later, those estrogen-bound receptors migrated to the nucleus, stimulating the transcription of various genes.[9,10] To date, several mechanisms for ER transcriptional regulation have been described in the literature. A classical mechanism is where E2 interacts with intracellular estrogen receptor (ERα or ERβ) resulting in receptor dimerization. This complex is then translocated to the nucleus, where it binds to estrogen response element (ERE) sequences through their DNA-binding domains.[11]

Estrogens are also shown to regulate transcription of several genes that do not contain EREs in their promoter regions, the mechanisms known as "indirect genomic signaling" or "transcriptional cross-talk." Here, estrogen indirect signaling influences activation or suppression of target gene expression by acting through protein–protein interactions with other transcription factors such as stimulating protein-1 (Sp-1), activating transcription factor (ATF)-2, Fos/c-jun, the ATF-1/cAMP (cyclic adenosine monophosphate) response element binding protein (ATF-1/CREB), and the nuclear transcription factor-Y (NF-Y).[12–16]

Yet, not all estrogen effects fit under the transcriptional regulation of steroid action. The observation of extremely fast estrogen-mediated biological responses led to the hypothesis that estrogen could be acting through mechanisms not involving the direct target gene of G Protein-Coupled Estrogen Receptor 1 (GPER1).[17] In addition, both ERα and ERβ have been identified outside the nucleus, *i.e.* in the cytoplasm, mitochondria, and associated with the plasma membrane,[18] from where they can rapidly activate other signaling cascades. Both the GPER1 and some variants of ERα or ERβ are associated with non-genomic estrogen signaling.[19,20] Non-genomic actions of the ERα or ERβ could be induced *via* a sub-population of receptors located at the cell membrane which activate intracellular signaling cascades such as the phospholipase C (PLC)/protein kinase C (PKCs) pathways,[21] the Ras/Raf/MAPK (mitogen-activated protein kinase) cascade,[22] the phosphatidyl inositol 3 kinase (PI3K)/Akt kinase cascade,[23] and the cAMP/protein kinase A (PKA) signaling pathway.[24,25]

ER can also be activated in the absence of estrogens or other receptor agonists, an interesting phenomenon observed in many cells and is known as "ligand-independent signaling".[26–28] This ligand-independent ER activation requires the action of regulatory molecules necessary for phosphorylation, such as PKA, PKC, MAPK phosphorylation cascade, inflammatory cytokines (interleukin (IL)-2), cell cycle regulators (RAS p21 protein activator cyclins A and D1), and peptide growth factors (epidermal growth factor (EGF), insulin, insulin-like growth factor-1, and transforming growth factor-β).[29]

These various mechanisms of action highlight the complex multifactorial processes induced by estrogen, estrogen-like molecules, and their cellular receptors. Several studies have shown the existence of additional convergent pathways involving both genomic and non-genomic factors that result in the regulation of gene transcription.[28,30]

1.3 Estrogenic Assessment of Xenoestrogens, and the Case of Bisphenol A

Xenoestrogens are man-made chemicals that disrupt the endocrine system by mimicking or interfering with the actions of estrogen. These disruptions can lead to estrogen dominance as well as developmental, reproductive, neurological, and immune effects. Xenoestrogens encompass a variety of chemicals, which may be of either synthetic or natural origin. Natural xenoestrogens are represented by phytoestrogens (derived from plants) and mycoestrogens (substances produced by fungi). Synthetic xenoestrogens are molecules produced by chemical synthesis, which are widely used in agricultural chemicals (pesticides) and industrial by-products (certain plastics or detergents), along with pharmaceutical estrogens.[31] Because of the ability of xenoestrogens to interfere with the endocrine system, they are also classified as endocrine-disrupting chemicals (EDCs). The estrogenic activity of some xenoestrogens such as octyl-phenol and bisphenol A (BPA) was accidentally discovered when they disrupted the experiments that studied the effects of natural estrogens.[32,33] Throughout the years, the appearance of adverse developmental and reproductive effects in aquatic and wildlife species living within or near areas contaminated with xenoestrogens was reported.[34–38] However, substantial evidence has pointed to the fact that these chemicals can mimic the action of the natural estrogen, although they do not exhibit a similar structure to that of estrogens. Thus, it became important to develop accurate assays that could evaluate the risk of xenoestrogens.

Commonly used *in vitro* screening methods are based on the classical concept of estrogenicity, such as competitive ER-binding assays and yeast-based reporter assays. However, as described earlier, E2 can mediate its estrogenic activity by many other signaling pathways, and not all of them are evaluated by these receptor-based assays. *In vitro* assays cannot detect pro-estrogens metabolized *in vivo* to estrogen, which further underestimates the potency of pro-estrogens and their metabolites.[39]

Bisphenol A, also known as BPA, is a synthetic man-made chemical with a molecular weight of 228.29 g mol^{-1} with the chemical formula $(CH_3)_2C(C_6H_4OH)_2$. This monomer was first synthesized by Dianin in 1891 and reported to be a synthetic estrogen in the 1930s.[40] In the 1950s, BPA was rediscovered as a compound that could be used to synthesize the first epoxy resins as protective coatings and later polymerized to make hard plastic called polycarbonate, which is strong enough to replace steel and clear enough to replace glass.[41] Since then, BPA has been used widely in industrial production and has become one of the highest volume chemicals produced worldwide. More than 6 billion pounds of BPA are produced each year, and >100 tons are released into the atmosphere by yearly production.[42]

Like other chemicals, BPA can be released (or leached) from these materials under heat stress or through acidic and basic conditions, which accelerates the hydrolysis of the ester bond linking BPA monomers and leads to human exposure.[43–45] It is estimated that BPA-contaminated food contributes to >90% of overall BPA exposure, whereas exposure through dental surgery, dermal absorption, and dust ingestion remains <5% in normal situations.[46] Overall, human exposure to BPA is consistent and widespread, and biomonitoring studies have reported that >90% of individuals have detectable amounts of BPA in urine samples in the United States, Germany, and Canada.[47–49] Exposure to BPA is a major health concern due to its ability to disrupt the endocrine system,[50,51] and, in many ways, it has become a model EDC. BPA has deleterious effects on the cardiovascular system, alters metabolism, contributes to cancer, and changes immune and reproductive systems. Exposure to BPA is associated with several human diseases.[52] In contrast with these reports, plastic manufacturers have started to release "BPA-free" plastic material, and the scientific community continues to report the risk of BPA for human beings and wildlife health, highlighting the demand for screening BPA exposures as a research priority.[53–55]

The dispute over the safety of the use of BPA has resulted in a deep divide between regulatory toxicologists working for federal agencies or chemical industries and scientists trained in the principles of endocrinology. These principles include the understanding of "low-dose" effects, non-monotonic dose responses, and co-exposure effects, as well as the presence of sex-specific and tissue-specific effects of BPA.[56–59] Another subject of significant debate surrounding BPA exposure is the possible mechanisms by which BPA is thought to exert endocrine-disrupting properties. BPA was initially thought to exert EDC actions primarily by disrupting the activity of the classical estrogen signaling pathways, using ERs as transcription factors binding to the ERE site in the DNA.[60,61] Nowadays, increasing BPA research shows that BPA can also trigger non-classical estrogen-activated pathways *via* binding to membrane ERs.[39,62] Moreover, many other signaling systems such as thyroid function,[63] androgen signaling,[64] hormone biosynthesis and/or metabolism and numerous other mechanisms that converge upon endocrine and reproductive systems have been proposed to explain BPA actions.[52,65] Far from addressing concerns about the safety of BPA doses accepted by government agencies, the incidence and prevalence of health problems associated with BPA exposure have increased worldwide. Numerous reviews are available regarding sources of BPA exposure, biomonitoring studies, mechanisms of action, and exposure effects, both *in vitro* and *in vivo*.[42,45] Furthermore, there are extensive reviews of BPA's low-dose effects and molecular mechanisms related to toxicity on human health.[51,52,65–67] Here, we examine the estrogenic activity of BPA when acting through the classical and non-classical ER-activated pathways to examine whether these mechanisms are at the root of the effect of BPA exposure on the endocrine system.

1.4 BPA-mediated Effects *via* Classical Estrogen Receptors

Like estrogens, BPA regulates different physiological processes such as growth, development, and homeostasis of numerous tissues through the binding and activation of the classical estrogen receptors, ERα and ERβ[51] (see Figure 1.1). These estrogen receptors are encoded by two separate genes located on human chromosomes 6 and 14, respectively.[68,69] In 1993, BPA's estrogenic activity was rediscovered while looking for an estrogen-binding protein in yeast. Krishnan *et al.* showed that BPA leached from polycarbonate flasks during autoclaving, rather than coming from yeast, as initially thought. Leaching was confirmed by performing different competitive binding assays such as ER and reversal estrogen action by tamoxifen, with the lowest effective dose being 10–20 nM.[32] The BPA molecule has structural features that confer the ability to bind to ERα and ERβ subtypes, although it displays 1000 to 10 000 times weak binding affinity to these receptors compared to the natural hormone E2.[35,65,70] The weaker affinity of BPA for ERs could be accounted for at least in part by the 42 van der Waals interactions within the BPA–ERα complex, in contrast to of the 51 interactions involved in the binding of E2–ERα.[71] Surprisingly, BPA could exert a stronger estrogen-like activity at nanomolar doses than at micromolar doses.[67,72]

Mechanisms for ER-mediated gene regulation are complex and depend on the recruitment of tissue-specific co-regulators that differentially affect ER interaction with EREs of different target genes.[73,74] BPA selectively binds to ERα and ERβ, but has a higher binding affinity relative to E2 at ERβ than at ERα in target cells.[75,76] A recent study in a rat model showed that BPA treatment decreased ERα and increased ERβ mRNA expression.[77] Similar effects were observed in female mice hearts exposed to 5 μg BPA/kg body weight during myocarditis, showing significant decreased ERα and increased ERβ mRNA expression levels.[78] Gould *et al.* previously emphasized that BPA is not merely a weak estrogen; rather, it exhibits characteristics of a distinct molecular mechanism of action *via* ERα.[79] An extensive uterotrophic analysis of BPA confirmed that it could induce proliferative and stimulatory changes in estrogen targets in concentrations ranging from 0.1 to 100 $mg\,kg^{-1}$.[80] Another study confirmed that BPA at a dose of 750 μg per mouse stimulated uterine proliferation in an ERα-dependent manner, which was highly correlated to that of E2.[81] In Ishikawa cells, BPA exposure decreased the protein expression

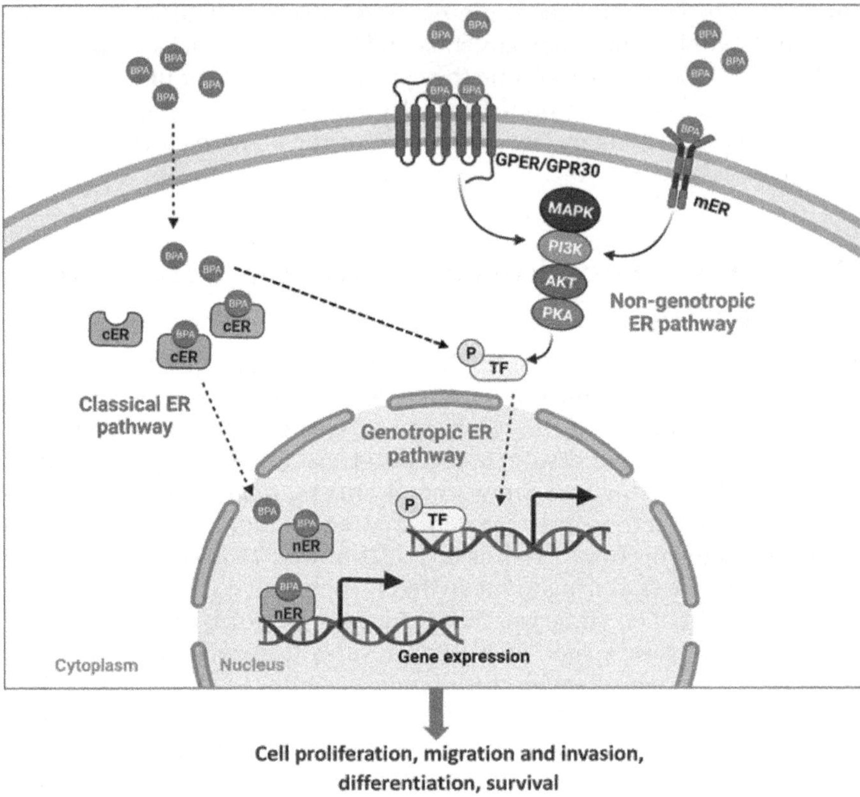

Figure 1.1 Mechanisms of bisphenol A (BPA) action *via* classical and non-classical estrogen receptor (ER) pathways. In the classical ER pathway, BPA binds with the ERs located in the cytoplasm (cER) or in the nucleus (nER) and regulates the transcription/translation of genes/proteins. In the non-genotropic pathway, BPA binds to membrane ER (mERs) and G-protein coupled ERs (GPER/GPR30), which triggers rapid estrogenic signaling *via* activation of cellular kinase systems such as mitogen-activated protein kinase (MAPK), phosphatidylinositol 3-kinase (PI3K), protein kinase A (PKA). In the non-classical genotropic ER pathway, BPA regulates gene expression by increasing or decreasing the expression of several transcription factors (TFs). P: phosphorylation. Created with BioRender.com.

levels of glucocorticoid-regulated kinase 1 (SGK1) and epithelial Na^+ channel α-subunit (ENaCα) *via* an ER pathway, causing damage in embryo implantation and uterus decidualization in mice.[82] During adulthood, BPA has been shown to bind with ERs located in the testis and causes harmful effects on germ cell differentiation, sperm production, and levels of male reproductive hormones.[83]

BPA induced the migration of normal human colon mucosal epithelial (NCM460) cells by increasing the expressions of integrin β1 and matrix metalloproteinase (MMP) mediated by ERβ.[84] In addition, BPA was shown to induce the key transcription factor of epithelial–mesenchymal transition (EMT) and alter the expression and nuclear localization of the zinc-finger transcription factor *Snai1* and in hemangioma cells *via* an $\$\mathrm{ER}\alpha$-mediated signaling pathway, resulting in the migration and invasion of hemangioma cells.[85] Such BPA-exposed induction of ER-dependent EMT was also observed in human breast cancer cells (MCF-7) by increasing the expression of C-X-C chemokine receptor type 4 (CXCR4), a receptor of CXCL12, thereby enhancing migration and invasion capacity of cells.[86] Up-regulation of CXCL12 by BPA exposure *via* ER activation was also reported in human ovarian carcinoma cells.[87] More recently, HepG2 cells exposed to BPA increased the transcriptional activity of *CYP2C9* by forming a BPA–ERα complex binding with ERE in the promoter region of *CYP2C9* gene.[88]

1.5 BPA-mediated Effects *via* Non-classical ER

Beyond the classical ER-modulator effect, BPA can also act *via* genomic and non-genomic pathways (see Figure 1.1). In the genomic pathway, BPA binds with the ERs located in the cytoplasm. The binding modifies overall receptor signaling that affects nuclear chromatin function and regulates the transcription and translation of genes and proteins, affecting cell proliferation, differentiation, and survival.[8] Studies have shown that BPA can facilitate estrogen-like activities similar to or stronger than E2.[89–91] These effects observed at low BPA doses can be explained at least partially by rapid responses *via* non-classical estrogen-triggered pathways.[39,52,65,66] BPA interacts differently within the ligand domain of ERs compared to E2.[79] BPA also has different co-activator recruitment, as highlighted by the fact that the BPA–ERb complex showed 500-fold greater potency than BPA–ERα in recruiting the co-activator TIF2.[76] Therefore, the adverse effects of BPA exposure on human health could be mediated by rapid activation of non-ER-dependent signaling pathways, which produce fast biological responses on specific cellular targets.

1.5.1 Non-genotropic ER Actions

Adverse effects of BPA exposure on human health are also mediated *via* non-genomic-dependent signaling pathways involving membrane estrogen receptor (mER).[92] Activation of mERs by BPA triggers rapid estrogenic signaling *via* activation of cellular kinase systems such as PKA, PKC, PI3K, MAPK, changes in levels of cAMP, and intracellular calcium as demonstrated in several human cancers.[93] High-dose BPA tends to acts as an ER antagonist by directly regulating the genomic transcription. In contrast, BPA at low doses is generally thought to disrupt the biological function in a non-genomic manner mediated by mERs.[94] BPA has been shown to promote cell proliferation or apoptosis by binding to mERs and rapid activation of downstream pathways.[95,96]

G-protein coupled estrogen receptor (GPER/GPR30) is one of the most common mERs that plays an important role in the toxic effects of BPA.[66,97] GPR30 mRNA is expressed in numerous tissues with different expression patterns.[98] GPR30 is a seven-transmembrane-domain receptor that binds BPA at a half maximal inhibitory concentration of 630 nM and its relative binding affinity when compared to E2 equals 2.83 nM in stably transfected ER-negative HEK293 cells.[92] Interestingly, studies have demonstrated that the binding affinity of E2 to GPR30 is 10-fold lower than ERα, whereas BPA affinity to GPR30 is about 50-fold higher than ERα.[92,99] In GC-2 cells, BPA-induced GPR30-mediated activation of the epidermal growth factor receptor (EGFR)-MAPK signaling pathway, along with activating the c-Fos gene and cell-cycle gene Cyclin D1 inhibition.[100] Similarly, BPA induced GPER-mediated activation of EGFR and MAPK signaling, leading to meiotic arrest in a zebrafish model.[101]

MAPK, PI3K/AKT, nuclear factor (NF)κB, JNK, and Ca^{2+} homeostasis are the most widely studied pathways associated with BPA and cancer.[93,102–104] BPA has been shown to activate the GPER1 (GPER/EGFR/extracellular signal-regulated kinase (ERK)1/2) signaling pathway in cancer cells by inducing the expression of the proto-oncogene c-fos (a significant gene in the early estrogen response) and activator protein-1 (AP1) target genes.[105,106] Similar signaling cascade activation was also reported by low dose of BPA, resulting in the progression of male germ-cell cancer.[107] In triple-negative breast cancer (TNBC) cells, BPA proliferative and pro-survival effects depend on ERK1/2 and AKT activation.[108] Constant BPA exposure to breast cancer cells induced resistance to an EGFR-targeted anti-cancer drug *via* EGFR/ERK1/2 pathway activation, and increased protein levels of BCL2 (anti-apoptotic) and SOD1 (anti-oxidant).[109,110] BPA-mediated resistance to anti-cancer drugs also activated pro-survival signaling pathways, such as PI3K/AKT/mTOR pathways.[111,112] Wang *et al.*[113] have demonstrated that BPA could stimulate the growth, invasion, and migration of RL95-2 cells *via* the MAPK pathway, which may upregulate cyclo-oxygenase-2 expression. BPA exposure was shown to induce migration and invasion of lung cancer cells and increased proliferation and migration of laryngeal squamous cell carcinoma (LSCC) cells mediated *via* the GPER-dependent pathway.[114,115] Interestingly, the BPA–GRP30 complex also induced testicular seminoma cell proliferation *in vitro*, the effect of which was reverted using a GPR30 antagonist, G15.[95] Castillo Sanchez *et al.*[116] found that BPA activates the GPER-dependent pathway and the kinases (*i.e.* focal adhesion kinase (FAK), proto-oncogene tyrosine-protein kinase (SRC), and ERK2) required for migration by increasing the activity of AP-1 and NFκB-DNA binding. In cervical cancer, BPA activated NFκB signaling *via* IKKβ resulting in cell migration and upregulation of fibronectin and metalloproteinase-9.[117] BPA also increased levels of leptin receptors, which induced proliferation by STAT3, ERK1/2, and AKT phosphorylation.[118] Additionally, BPA-induced Ca^{2+} signaling initiated at the plasma membrane activated the transcription factor CREB in the nucleus within 15 minutes of exposure.[119] Yaguchi[120] reported that BPA activated the EGFR/ERK pathway by facilitating Ca^{2+} influx in Ishikawa cells and EGF secretion to the extracellular space. BPA also induced apoptosis of KGN cells (a human granulosa-like tumor cell line) through GPER-dependent activation of the ROS/Ca^{2+}–ASK1–JNK signaling pathway.[121] BPA was shown to impair glucose tolerance, increase body weight, and reduce insulin secretion in mice,[122,123] effects that were protected in GPR30 knockout female mice.[124] Exposure to low-dose BPA increases GPR30 and produces specific inflammatory proteins, including IL8, IL6, and monocyte chemoattractant protein-1α, both in cultured mature adipocytes and in stromal-vascular fraction cells isolated from mammary human adipose tissue biopsies.[125]

Another membrane-associated ER, named ERα36, was reported to promote BPA-induced leiomyoma cell proliferation by activating SRC, EGFR, and MAPK p44/42, increase expression of growth factor receptor-bound protein 2 (*Grb2*), son of sevenless homolog 1 (*Sos1*) and Ras.[126] Integrins are transmembrane receptors that can regulate cellular signals. BPA was found to increase the expressions of MMP-2, MMP-9, integrin β1, and integrin α5 in the placenta, inducing a greater proportion of the labyrinth and spongiotrophoblast layers in mice.[127] Environmental doses of BPA induced cancer cell migration *via* a rapid direct activation of integrin β1.[128]

1.5.2 Genotropic ER Actions

Several nuclear transcription factors (TFs) are associated with BPA action *via* non-classical ER signaling (see Figure 1.1). In particular, studies have shown that induction of adipogenic TFs, such as PPARγ, C/EBPs, and Nrf2 play an important role in the "obesogenic effect" of BPA. Other evidence suggests a critical role of HAND2 protein and HOX family members in BPA-mediated detrimental effects.

1.5.2.1 PPARγ

Peroxisome proliferator-activated receptors (PPARs) are members of the nuclear receptor superfamily known to exert a wide range of biological effects on adipogenesis, cell proliferation, differentiation, immune response, and metabolism. The PPAR family comprises three different subtypes (PPARα, β/δ, γ), all of which are important regulators of glucose and lipid metabolism.[52] The activity of PPARγ is governed by the binding of small lipophilic ligands, chiefly fatty acids, derived from nutrition or metabolism. Perinatal BPA exposure *via* milk during lactation modified early adipogenesis by modulating adipocyte hypertrophy and overexpression of pro-adipogenic transcription factors, PPARγ, increasing the body weight of female pups, but not male pups.[129] Similarly, gestational BPA exposure increases PPARγ expression level in pre-adipocytes isolated from female sheep progeny, but not from male progeny.[130]

It has been reported that BPA effects mediated *via* PPARγ are not only observed in adipose tissue, but also in the liver. Subcutaneous injection of BPA in mice increases the PPARγ gene expression levels in the liver, resulting in fasting hyperglycemia, glucose intolerance, and high levels of non-esterified fatty acids.[122] Hepatic PPARγ expression, along with increased fat mass and total body weight, were also observed in male mice exposed to BPA.[131] Several conflicting data on PPARγ-mediated BPA effects have been obtained in culture cells. Increased PPARγ levels were observed in murine 3T3-L1 cells exposed to both low and high doses of BPA.[131,132] Conversely, no differences in PPARγ 1 and PPARγ 2 expression levels were observed in the same cells in response to BPA exposure.[133]

BPA has also been shown to significantly upregulate PPARγ expression in adult human pre-adipocytes and freshly cultured omental adipose tissue from child donors.[134,135] In contrast, BPA effects in human adipose-derived stem cells was not shown to be mediated by PPARγ.[134]

1.5.2.2 C/EBP

Similar to that of PPARγ, the role of CCAAT/enhancer-binding proteins (C/EBPs) as mediators of BPA effects is presently disputed. C/EBPs encompasses a family of six transcription factors with structural and functional homologies, but with different transactivating abilities and tissue specificities.

While BPA-exposed female rats have reported an increase of C/EBP expression in adipocytes,[129] no effect of BPA was observed on this transcription factor in 3T3-L1 cells.[133] A recent study highlighted the involvement of other members of the C/EBP family in the metabolic damage caused by BPA, where triglyceride accumulation in human adipose-derived mesenchymal stem cells by BPA exposure is associated not only to the upregulation of PPARγ and C/EBPα, but also to the increase of C/EBPβ gene expression.[136] C/EBP-mediated BPA effects have also been linked with liver dysfunction and disease. Female mice fed a diet supplemented with BPA showed a decreased level of C/EBPα in fetal livers compared to the control animals, an effect not observed in male mice. This highlights BPA-mediated disruption of fetal liver maturation in a sex-specific manner and may further alter the expression level of albumin, alpha-fetoprotein, and glycogen synthase.[137]

1.5.2.3 Nrf2

Nuclear factor eythroid-2-related factor 2 (Nrf2) is a basic leucine zipper transcription factor regulating the expression of antioxidant proteins and protects against oxidative damage.[138] BPA administered orally to lupus-prone MRL/lpr mice has been shown to decrease Nrf2 expression in renal tissue exacerbating lupus nephritis, highlighting a protective role of Nrf2 in BPA-induced renal damage.[139] Unlike the kidneys, Nrf2 impairs liver function in leptin-deficient mice by decreasing Kelch-like ECH-associated protein 1 (Keap1) and increasing lipid accumulation associated with constitutive activation of Nrf2.[140] Such BPA-induced Nrf2 upregulation *via* Keap1 inactivation has been reported in a human hepatoma cell line.[141] Furthermore, 25 $\mu g\,kg^{-1}$ per day of BPA administration to pregnant CD-1 mice increased Nrf2 expression and its recruitment to the Srebp-1c promoter, causing hepatic lipid deposition.[142]

1.5.2.4 HAND2

Heart- and neural crest derivatives-expressed protein 2 (HAND2) is a basic helix-loop-helix transcription factor, which plays an important role in establishing proper implantation for pregnancy. Chronic BPA exposure to female mice decreases HAND2 expression in the uterine stroma, affecting embryo implantation and formation of the decidua during the early phases of pregnancy.[143] HAND2 overexpression is also associated with increased proliferation of cardiac progenitor cells,[144] and emerging evidence indicates an association between cardiovascular diseases and BPA.[145] Interestingly, BPA significantly upregulates ERβ expression and H3K9 and H4K12 histone acetylation, which could be responsible for HAND2 upregulation and the increased percentage of heart malformations.[146] Thus, HAND2 represents a key regulator of several organs such as the uterus and heart, and impairment of its expression following BPA exposure may lead to reproductive and cardiac disorders *via* genetic and epigenetic mechanisms.

1.5.2.5 HOX

The term "HOX" was originally termed as a group of related HOX genes that encode for transcription factors characterized by a well-conserved DNA sequence known as the homeobox. HOX genes are found in humans and rodents and are expressed during embryogenesis and early development, where they act as master transcriptional regulators.[147,148] Among the HOX genes, HOXA10 is expressed in the female reproductive system and is important for normal decidualization and pregnancy, while HOXB9 and HOXC6 are involved in the development of the mammary gland.[149–151] Female pups exposed in utero to 0.5–1.0 $mg\,kg^{-1}$ BPA showed an elevated level of HOXA10 in uterine stromal cells, which may mediate the decidualization defects.[151] HOXB9 is also known to activate in response to BPA exposure in breast cancer cells.[102] Studies have also shown that BPA increases HOXB9 and HOXC6 expression in the mammary glands of ovariectomized rats and culture human breast cancer cells (MCF7), highlighting that BPA harmful effects in breast tumors are mediated *via* these transcription factors.[152,153]

1.6 Conclusion

Most of the world's population is widely exposed to BPA, as it is largely used to manufacture polycarbonate plastic and is released into foods and beverages. A growing body of evidence indicates that BPA exposure has been linked to numerous adverse perinatal, childhood, and adult health outcomes in both animals and humans, such as reproductive effects, developmental effects, metabolic disease, and cancer.[93,154,155] However, several contrasting results about the adverse effects of BPA have been described, which may be due to the use of different doses, experimental models or conditions.[42] In parallel, increasing efforts have been taken to elucidate the molecular mechanisms through which BPA acts. The integration of the knowledge about the molecular pathways of BPA with epidemiology could certainly improve the understanding of the toxic effects of BPA on human health.

As summarized in this chapter, there is no doubt that BPA is an estrogenic compound that can initiate the classical ER pathway through binding to relatively specific ERs and regulating gene expression. Additionally, BPA-estrogenic action also acts through non-classical estrogen-activated pathways. Rapid non-genomic and genomic actions may be mediated by BPA's interactions with the membrane-associated ERs and/or GPERs and a suite of important transcription factors. BPA exposure significantly impacts growth, survival, proliferation, invasion, migration, and apoptosis in cell- and tissue-specific manners through these mechanisms. In addition, exposure to BPA may facilitate chemotherapy resistance to anti-cancer drugs and also interacts with several receptors and causes aberrant changes in numerous pathways such as GPER/EGFR/ERK1/2, JAK/STAT, PI3K/AKT/mTOR, SRC1-3, phosphorylation of cyclin D1, AKT, PPARγ, and Ca^{2+} homeostasis. BPA probably exerts these adverse effects with rapid signaling, showing the presence of extracellularly accessible binding sites that act to regulate intracellular signaling. This rapid signaling may elicit signal transduction pathways responsible for growth and differentiation and in energy and nutrient metabolism. It has become evident that BPA can activate transduction signaling pathways that vary across cell types, and the total disruption effect arises from the combination of rapid mechanisms and longer signaling effects, as highlighted in this chapter. The mechanisms at the root of these multiple effects are numerous and are activated at BPA concentrations below the concentration range in which "pharmaceutical" effects are detected by classical toxicology and create non-monotonic dose–response curves.[156] It is also interesting to note that all the upstream pathways may contribute to stable and inheritable modifications by regulating epigenetic enzymes, which may also sustain initial exposure to BPA.[157] Nevertheless, these mechanisms are only applicable and reported in some cell types and animal models. More mechanistic and comprehensive insights into other biological systems are needed to unravel the classical and non-classical ER effects of BPA. Identifying BPA effects in different model systems may help raise awareness within the scientific community and the manufacturing industry of the need to seek alternatives to BPA to reduce its harmful effects. Therefore, to date, the best practice is still the precaution of limiting the use of plastic materials and promoting BPA-free products.

Acknowledgements

I thank Dr Natalie R. Gassman (University of South Alabama, Mobile, AL, USA) for critically reading the chapter and suggesting substantial improvements.

References

1. J. Liang and Y. Shang, Estrogen and cancer, *Annu. Rev. Physiol.*, 2013, **75**, 225–240.
2. E. R. Simpson, M. Misso, K. N. Hewitt, R. A. Hill, W. C. Boon, M. E. Jones, A. Kovacic, J. Zhou and C. D. Clyne, Estrogen–the good, the bad, and the unexpected, *Endocr. Rev.*, 2005, **26**, 322–330.
3. C. J. Gruber, W. Tschugguel, C. Schneeberger and J. C. Huber, Production and actions of estrogens, *N. Engl. J. Med.*, 2002, **346**, 340–352.
4. R. D. Koos, Minireview: Putting physiology back into estrogens' mechanism of action, *Endocrinology*, 2011, **152**, 4481–4488.
5. J. L. Voogt, Control of hormone release during lactation, *Clin. Obstet. Gynecol.*, 1978, **5**, 435–455.
6. M. Schulster, A. M. Bernie and R. Ramasamy, The role of estradiol in male reproductive function, *Asian J. Androl.*, 2016, **18**, 435–440.
7. J.-Å. Gustafsson, What pharmacologists can learn from recent advances in estrogen signalling, *Trends Pharmacol. Sci.*, 2003, **24**, 479–485.
8. N. Heldring, A. Pike, S. Andersson, J. Matthews, G. Cheng, J. Hartman, M. Tujague, A. Ström, E. Treuter, M. Warner and J.-Å. Gustafsson, Estrogen Receptors: How Do They Signal and What Are Their Targets, *Physiol. Rev.*, 2007, **87**, 905–931.
9. E. V. Jensen, E. R. Desombre, T. Kawashima, T. Suzuki, K. Kyser and P. W. Jungblut, Estrogen-binding substances of target tissues, *Science*, 1967, **158**, 529–530.
10. E. V. Jensen, T. Suzuki, T. Kawashima, W. E. Stumpf, P. W. Jungblut and E. R. DeSombre, A two-step mechanism for the interaction of estradiol with rat uterus, *Proc. Natl. Acad. Sci. U. S. A.*, 1968, **59**, 632–638.
11. N. Fuentes and P. Silveyra, Estrogen receptor signaling mechanisms, *Adv. Protein Chem. Struct. Biol.*, 2019, **116**, 135–170.
12. A. Aranda and A. Pascual, Nuclear hormone receptors and gene expression, *Physiol. Rev.*, 2001, **81**, 1269–1304.
13. M. Göttlicher, S. Heck and P. Herrlich, Transcriptional cross-talk, the second mode of steroid hormone receptor action, *J. Mol. Med.*, 1998, **76**, 480–489.
14. R. O'Lone, M. C. Frith, E. K. Karlsson and U. Hansen, Genomic targets of nuclear estrogen receptors, *Mol. Endocrinol.*, 2004, **18**, 1859–1875.
15. S. Safe and K. Kim, Nuclear receptor-mediated transactivation through interaction with Sp proteins, *Prog. Nucleic Acid Res. Mol. Biol.*, 2004, 77, 1–36.
16. B. Saville, M. Wormke, F. Wang, T. Nguyen, E. Enmark, G. Kuiper, J. A. Gustafsson and S. Safe, Ligand-, cell-, and estrogen receptor

subtype (alpha/beta)-dependent activation at GC-rich (Sp1) promoter elements, *J. Biol. Chem.*, 2000, **275**, 5379–5387.
17. E. R. Prossnitz and M. Barton, The G-protein-coupled estrogen receptor GPER in health and disease, *Nat. Rev. Endocrinol.*, 2011, **7**, 715–726.
18. S. R. Hammes and E. R. Levin, Extranuclear steroid receptors: nature and actions, *Endocr. Rev.*, 2007, **28**, 726–741.
19. M. Barton, E. J. Filardo, S. J. Lolait, P. Thomas, M. Maggiolini and E. R. Prossnitz, Twenty years of the G protein-coupled estrogen receptor GPER: Historical and personal perspectives, *J. Steroid Biochem. Mol. Biol.*, 2018, **176**, 4–15.
20. E. J. Filardo and P. Thomas, Minireview: G protein-coupled estrogen receptor-1, GPER-1: its mechanism of action and role in female reproductive cancer, renal and vascular physiology, *Endocrinology*, 2012, **153**, 2953–2962.
21. M. J. Marino, S. T. Rouse, A. I. Levey, L. T. Potter and P. J. Conn, Activation of the genetically defined m1 muscarinic receptor potentiates N-methyl-D-aspartate (NMDA) receptor currents in hippocampal pyramidal cells, *Proc. Natl. Acad. Sci. U. S. A.*, 1998, **95**, 11465–11470.
22. E. G. Dos Santos, M. N. Dieudonne, R. Pecquery, V. Le Moal, Y. Giudicelli and D. Lacasa, Rapid nongenomic E2 effects on p42/p44 MAPK, activator protein-1, and cAMP response element binding protein in rat white adipocytes, *Endocrinology*, 2002, **143**, 930–940.
23. M. Marino, F. Acconcia and A. Trentalance, Biphasic estradiol-induced AKT phosphorylation is modulated by PTEN via MAP kinase in HepG2 cells, *Mol. Biol. Cell*, 2003, **14**, 2583–2591.
24. Q. Gu and R. L. Moss, 17 beta-Estradiol potentiates kainate-induced currents via activation of the cAMP cascade, *J. Neurosci.*, 1996, **16**, 3620–3629.
25. G. Picotto, V. Massheimer and R. Boland, Acute stimulation of intestinal cell calcium influx induced by 17β-estradiol via the cAMP messenger system, *Mol. Cell. Endocrinol.*, 1996, **119**, 129–134.
26. M. A. Bennesch and D. Picard, Minireview: Tipping the balance: ligand-independent activation of steroid receptors, *Mol. Endocrinol.*, 2015, **29**, 349–363.
27. A. Maggi, Liganded and unliganded activation of estrogen receptor and hormone replacement therapies, *Biochim Biophys Acta*, 2011, **1812**, 1054–1060.
28. P. Vrtačnik, B. Ostanek, S. Mencej-Bedrač and J. Marc, The many faces of estrogen signaling, *Biochem. Med.*, 2014, **24**, 329–342.
29. S. Nilsson, S. Mäkelä, E. Treuter, M. Tujague, J. Thomsen, G. Andersson, E. Enmark, K. Pettersson, M. Warner and J. A. Gustafsson, Mechanisms of estrogen action, *Physiol. Rev.*, 2001, **81**, 1535–1565.
30. L. Björnström and M. Sjöberg, Mechanisms of estrogen receptor signaling: convergence of genomic and nongenomic actions on target genes, *Mol. Endocrinol.*, 2005, **19**, 833–842.

31. I. Paterni, C. Granchi and F. Minutolo, Risks and benefits related to alimentary exposure to xenoestrogens, *Crit. Rev. Food Sci. Nutr.*, 2017, **57**, 3384–3404.
32. A. V. Krishnan, P. Stathis, S. F. Permuth, L. Tokes and D. Feldman, Bisphenol-A: an estrogenic substance is released from polycarbonate flasks during autoclaving, *Endocrinology*, 1993, **132**, 2279–2286.
33. A. M. Soto, H. Justicia, J. W. Wray and C. Sonnenschein, *p*-Nonylphenol: an estrogenic xenobiotic released from "modified" polystyrene, *Environ. Health Perspect.*, 1991, **92**, 167–173.
34. L. J. Guillette, Jr., T. S. Gross, G. R. Masson, J. M. Matter, H. F. Percival and A. R. Woodward, Developmental abnormalities of the gonad and abnormal sex hormone concentrations in juvenile alligators from contaminated and control lakes in Florida, *Environ. Health Perspect.*, 1994, **102**, 680–688.
35. M. Sonavane, N. Creusot, E. Maillot-Maréchal, A. Péry, F. Brion and S. Aït-Aïssa, Zebrafish-based reporter gene assays reveal different estrogenic activities in river waters compared to a conventional human-derived assay, *Sci. Total Environ.*, 2016, **550**, 934–939.
36. M. Sonavane, J. E. Schollée, A. O. Hidasi, N. Creusot, F. Brion, M. J. Suter, J. Hollender and S. Aït-Aïssa, An integrative approach combining passive sampling, bioassays, and effect-directed analysis to assess the impact of wastewater effluent, *Environ. Toxicol. Chem.*, 2018, **37**, 2079–2088.
37. C. Sonnenschein and A. M. Soto, *The Society of Cells – Cancer and Control of Cell Proliferation*, Bios Scientific, Oxford, 1999, 154.
38. J. P. Sumpter and S. Jobling, Male sexual development in "a sea of oestrogen", *Lancet*, 1993, **342**, 124–125.
39. P. Alonso-Magdalena, A. B. Ropero, S. Soriano, M. García-Arévalo, C. Ripoll, E. Fuentes, I. Quesada and Á. Nadal, Bisphenol-A acts as a potent estrogen via non-classical estrogen triggered pathways, *Mol. Cell. Endocrinol.*, 2012, **355**, 201–207.
40. E. C. Dodds and W. Lawson, Synthetic strogenic Agents without the Phenanthrene Nucleus, *Nature*, 1936, **137**, 996.
41. S. A. Vogel, The politics of plastics: the making and unmaking of bisphenol a "safety", *Am. J. Public Health*, 2009, **99**(Suppl 3), S559–S566.
42. L. N. Vandenberg, M. V. Maffini, C. Sonnenschein, B. S. Rubin and A. M. Soto, Bisphenol-A and the great divide: a review of controversies in the field of endocrine disruption, *Endocr. Rev.*, 2009, **30**, 75–95.
43. A. M. Calafat, Z. Kuklenyik, J. A. Reidy, S. P. Caudill, J. Ekong and L. L. Needham, Urinary concentrations of bisphenol A and 4-nonylphenol in a human reference population, *Environ. Health Perspect.*, 2005, **113**, 391–395.
44. J. H. Kang, F. Kondo and Y. Katayama, Human exposure to bisphenol A, *Toxicology*, 2006, **226**, 79–89.

45. C. A. Richter, L. S. Birnbaum, F. Farabollini, R. R. Newbold, B. S. Rubin, C. E. Talsness, J. G. Vandenbergh, D. R. Walser-Kuntz and F. S. vom Saal, In vivo effects of bisphenol A in laboratory rodent studies, *Reprod. Toxicol.*, 2007, **24**, 199–224.
46. T. Geens, D. Aerts, C. Berthot, J. P. Bourguignon, L. Goeyens, P. Lecomte, G. Maghuin-Rogister, A. M. Pironnet, L. Pussemier, M. L. Scippo, J. Van Loco and A. Covaci, A review of dietary and non-dietary exposure to bisphenol-A, *Food Chem. Toxicol.*, 2012, **50**, 3725–3740.
47. T. Bushnik, D. Haines, P. Levallois, J. Levesque, J. Van Oostdam and C. Viau, Lead and bisphenol A concentrations in the Canadian population, *Health Rep.*, 2010, **21**, 7–18.
48. H. M. Koch, M. Kolossa-Gehring, C. Schröter-Kermani, J. Angerer and T. Brüning, Bisphenol A in 24 h urine and plasma samples of the German Environmental Specimen Bank from 1995 to 2009: a retrospective exposure evaluation, *J. Exposure Sci. Environ. Epidemiol.*, 2012, **22**, 610–616.
49. J. S. LaKind, J. Levesque, P. Dumas, S. Bryan, J. Clarke and D. Q. Naiman, Comparing United States and Canadian population exposures from National Biomonitoring Surveys: Bisphenol A intake as a case study, *J. Exposure Sci. Environ. Epidemiol.*, 2012, **22**, 219–226.
50. B. S. Rubin, Bisphenol A: an endocrine disruptor with widespread exposure and multiple effects, *J. Steroid Biochem. Mol. Biol.*, 2011, **127**, 27–34.
51. Y. B. Wetherill, B. T. Akingbemi, J. Kanno, J. A. McLachlan, A. Nadal, C. Sonnenschein, C. S. Watson, R. T. Zoeller and S. M. Belcher, In vitro molecular mechanisms of bisphenol A action, *Reprod. Toxicol.*, 2007, **24**, 178–198.
52. I. Cimmino, F. Fiory, G. Perruolo, C. Miele, F. Beguinot, P. Formisano and F. Oriente, Potential Mechanisms of Bisphenol A (BPA) Contributing to Human Disease, *Int. J. Mol. Sci.*, 2020, **21**, 5761.
53. A. Bergman, J. J. Heindel, T. Kasten, K. A. Kidd, S. Jobling, M. Neira, R. T. Zoeller, G. Becher, P. Bjerregaard, R. Bornman, I. Brandt, A. Kortenkamp, D. Muir, M.-N. B. Drisse, R. Ochieng, N. E. Skakkebaek, A. S. Byléhn, T. Iguchi, J. Toppari and T. J. Woodruff, The impact of endocrine disruption: a consensus statement on the state of the science, *Environ. Health Perspect.*, 2013, **121**, A104–A106.
54. V. Le Fol, S. Aït-Aïssa, M. Sonavane, J. M. Porcher, P. Balaguer, J. P. Cravedi, D. Zalko and F. Brion, In vitro and in vivo estrogenic activity of BPA, BPF and BPS in zebrafish-specific assays, *Ecotoxicol. Environ. Saf.*, 2017, **142**, 150–156.
55. R. T. Zoeller, T. R. Brown, L. L. Doan, A. C. Gore, N. E. Skakkebaek, A. M. Soto, T. J. Woodruff and F. S. Vom Saal, Endocrine-disrupting chemicals and public health protection: a statement of principles from The Endocrine Society, *Endocrinology*, 2012, **153**, 4097–4110.
56. B. F. Healy, K. R. English, P. Jagals and P. D. Sly, Bisphenol A exposure pathways in early childhood: Reviewing the need for improved risk

assessment models, *J. Exposure Sci. Environ. Epidemiol.*, 2015, **25**, 544–556.

57. M. Sonavane and N. R. Gassman, Bisphenol A co-exposure effects: a key factor in understanding BPA's complex mechanism and health outcomes, *Crit. Rev. Toxicol.*, 2019, **49**, 371–386.
58. R. Valentino, V. D'Esposito, F. Ariemma, I. Cimmino, F. Beguinot and P. Formisano, Bisphenol A environmental exposure and the detrimental effects on human metabolic health: is it necessary to revise the risk assessment in vulnerable population?, *J. Endocrinol. Invest.*, 2016, **39**, 259–263.
59. F. S. vom Saal and C. Hughes, An extensive new literature concerning low-dose effects of bisphenol A shows the need for a new risk assessment, *Environ. Health Perspect.*, 2005, **113**, 926–933.
60. G. G. Kuiper, J. G. Lemmen, B. Carlsson, J. C. Corton, S. H. Safe, P. T. van der Saag, B. van der Burg and J. A. Gustafsson, Interaction of estrogenic chemicals and phytoestrogens with estrogen receptor beta, *Endocrinology*, 1998, **139**, 4252–4263.
61. C. Q. Sheeler, M. W. Dudley and S. A. Khan, Environmental estrogens induce transcriptionally active estrogen receptor dimers in yeast: activity potentiated by the coactivator RIP140, *Environ. Health Perspect.*, 2000, **108**, 97–103.
62. A. L. Wozniak, N. N. Bulayeva and C. S. Watson, Xenoestrogens at picomolar to nanomolar concentrations trigger membrane estrogen receptor-alpha-mediated Ca^{2+} fluxes and prolactin release in GH3/B6 pituitary tumor cells, *Environ. Health Perspect.*, 2005, **113**, 431–439.
63. R. T. Zoeller, Environmental chemicals as thyroid hormone analogues: new studies indicate that thyroid hormone receptors are targets of industrial chemicals?, *Mol. Cell. Endocrinol.*, 2005, **242**, 10–15.
64. H. J. Lee, S. Chattopadhyay, E.-Y. Gong, R. S. Ahn and K. Lee, Antiandrogenic Effects of Bisphenol A and Nonylphenol on the Function of Androgen Receptor, *Toxicol. Sci.*, 2003, **75**, 40–46.
65. F. Acconcia, V. Pallottini and M. Marino, Molecular Mechanisms of Action of BPA, *Dose-Response*, 2015, **13**, 1559325815610582.
66. Y. Ma, H. Liu, J. Wu, L. Yuan, Y. Wang, X. Du, R. Wang, P. W. Marwa, P. Petlulu, X. Chen and H. Zhang, The adverse health effects of bisphenol A and related toxicity mechanisms, *Environ. Res.*, 2019, **176**, 108575.
67. F. S. Vom Saal and L. N. Vandenberg, Update on the Health Effects of Bisphenol A: Overwhelming Evidence of Harm, *Endocrinology*, 2021, **162**, bqaa171.
68. E. Enmark, M. Pelto-Huikko, K. Grandien, S. Lagercrantz, J. Lagercrantz, G. Fried, M. Nordenskjöld and J. A. Gustafsson, Human estrogen receptor beta-gene structure, chromosomal localization, and expression pattern, *J. Clin. Endocrinol. Metab.*, 1997, **82**, 4258–4265.
69. L. P. Menasce, G. R. White, C. J. Harrison and J. M. Boyle, Localization of the estrogen receptor locus (ESR) to chromosome 6q25.1 by

FISH and a simple post-FISH banding technique, *Genomics*, 1993, **17**, 263–265.
70. H. Fang, W. Tong, R. Perkins, A. M. Soto, N. V. Prechtl and D. M. Sheehan, Quantitative comparisons of in vitro assays for estrogenic activities, *Environ. Health Perspect.*, 2000, **108**, 723–729.
71. V. Delfosse, M. Grimaldi, J. L. Pons, A. Boulahtouf, A. le Maire, V. Cavailles, G. Labesse, W. Bourguet and P. Balaguer, Structural and mechanistic insights into bisphenols action provide guidelines for risk assessment and discovery of bisphenol A substitutes, *Proc. Natl. Acad. Sci. U. S. A.*, 2012, **109**, 14930–14935.
72. W. V. Welshons, K. A. Thayer, B. M. Judy, J. A. Taylor, E. M. Curran and F. S. Vom Saal, Large effects from small exposures. I. Mechanisms for endocrine-disrupting chemicals with estrogenic activity, *Environ. Health Perspect.*, 2003, **111**, 994–1006.
73. C. M. Klinge, Estrogen receptor interaction with co-activators and co-repressors, *Steroids*, 2000, **65**, 227–251.
74. C. M. Klinge, Estrogen receptor interaction with estrogen response elements, *Nucleic Acids Res.*, 2001, **29**, 2905–2919.
75. G. G. Kuiper, B. Carlsson, K. Grandien, E. Enmark, J. Häggblad, S. Nilsson and J. A. Gustafsson, Comparison of the ligand binding specificity and transcript tissue distribution of estrogen receptors alpha and beta, *Endocrinology*, 1997, **138**, 863–870.
76. E. J. Routledge, R. White, M. G. Parker and J. P. Sumpter, Differential effects of xenoestrogens on coactivator recruitment by estrogen receptor (ER) alpha and ERbeta, *J. Biol. Chem.*, 2000, **275**, 35986–35993.
77. S. S. Mohamad Zaid, N. M. Kassim and S. Othman, Tualang Honey Protects against BPA-Induced Morphological Abnormalities and Disruption of ERα, ERβ, and C3 mRNA and Protein Expressions in the Uterus of Rats, *Evidence-Based Complementary Altern. Med.*, 2015, **2015**, 202874.
78. K. A. Bruno, J. E. Mathews, A. L. Yang, J. A. Frisancho, A. J. Scott, H. D. Greyner, F. A. Molina, M. S. Greenaway, G. M. Cooper, A. Bucek, A. C. Morales-Lara, A. R. Hill, A. A. Mease, D. N. Di Florio, J. M. Sousou, A. C. Coronado, A. R. Stafford and D. Fairweather, BPA Alters Estrogen Receptor Expression in the Heart After Viral Infection Activating Cardiac Mast Cells and T Cells Leading to Perimyocarditis and Fibrosis, *Front. Endocrinol.*, 2019, **10**, 598.
79. J. C. Gould, L. S. Leonard, S. C. Maness, B. L. Wagner, K. Conner, T. Zacharewski, S. Safe, D. P. McDonnell and K. W. Gaido, Bisphenol A interacts with the estrogen receptor alpha in a distinct manner from estradiol, *Mol. Cell. Endocrinol.*, 1998, **142**, 203–214.
80. C. M. Markey, E. H. Luque, M. Munoz De Toro, C. Sonnenschein and A. M. Soto, In utero exposure to bisphenol A alters the development and tissue organization of the mouse mammary gland, *Biol. Reprod.*, 2001, **65**, 1215–1223.

81. S. C. Hewitt and K. S. Korach, Estrogenic activity of bisphenol A and 2,2-bis(*p*-hydroxyphenyl)-1,1,1-trichloroethane (HPTE) demonstrated in mouse uterine gene profiles, *Environ. Health Perspect.*, 2011, **119**, 63–70.
82. M. Yuan, M. Hu, Y. Lou, Q. Wang, L. Mao, Q. Zhan and F. Jin, Environmentally relevant levels of bisphenol A affect uterine decidualization and embryo implantation through the estrogen receptor/serum and glucocorticoid-regulated kinase 1/epithelial sodium ion channel α-subunit pathway in a mouse model, *Fertil. Steril.*, 2018, **109**, 735–744, .e731.
83. C. Williams, M. Bondesson, D. N. Krementsov and C. Teuscher, Gestational bisphenol A exposure and testis development, *Endocr. Disruptors*, 2014, **2**, e29088.
84. T. Shi, C. Zhao, Z. Li, Q. Zhang and X. Jin, Bisphenol a exposure promotes the migration of NCM460 cells via estrogen receptor-mediated integrin β1/MMP-9 pathway, *Environ. Toxicol.*, 2016, **31**, 799–807.
85. D. Zhai, J. He, X. Li, L. Gong and Y. Ouyang, Bisphenol A regulates Snail-mediated epithelial-mesenchymal transition in hemangioma cells, *Cell Biochem. Funct.*, 2016, **34**, 441–448.
86. G. A. Lee, K. A. Hwang and K. C. Choi, Inhibitory effects of 3,3′-diindolylmethane on epithelial-mesenchymal transition induced by endocrine disrupting chemicals in cellular and xenograft mouse models of breast cancer, *Food Chem. Toxicol.*, 2017, **109**, 284–295.
87. J. M. Hall and K. S. Korach, Endocrine disrupting chemicals promote the growth of ovarian cancer cells via the ER-CXCL12-CXCR4 signaling axis, *Mol. Carcinog.*, 2013, **52**, 715–725.
88. J.-Y. Xu, L. Wu, Z. Shi, X.-J. Zhang, N. A. Englert and S.-Y. Zhang, Up-regulation of human CYP2C9 expression by Bisphenol A via estrogen receptor alpha (ERα) and Med25, *Environ. Toxicol.*, 2017, **32**, 970–978.
89. P. Alonso-Magdalena, S. Morimoto, C. Ripoll, E. Fuentes and A. Nadal, The estrogenic effect of bisphenol A disrupts pancreatic beta-cell function in vivo and induces insulin resistance, *Environ. Health Perspect.*, 2006, **114**, 106–112.
90. E. R. Hugo, T. D. Brandebourg, J. G. Woo, J. Loftus, J. W. Alexander and N. Ben-Jonathan, Bisphenol A at environmentally relevant doses inhibits adiponectin release from human adipose tissue explants and adipocytes, *Environ. Health Perspect.*, 2008, **116**, 1642–1647.
91. A. Zsarnovszky, H. H. Le, H. S. Wang and S. M. Belcher, Ontogeny of rapid estrogen-mediated extracellular signal-regulated kinase signaling in the rat cerebellar cortex: potent nongenomic agonist and endocrine disrupting activity of the xenoestrogen bisphenol A, *Endocrinology*, 2005, **146**, 5388–5396.
92. P. Thomas and J. Dong, Binding and activation of the seven-transmembrane estrogen receptor GPR30 by environmental estrogens: a potential novel mechanism of endocrine disruption, *J. Steroid Biochem. Mol. Biol.*, 2006, **102**, 175–179.

93. N. G. Khan, J. Correia, D. Adiga, P. S. Rai, H. S. Dsouza, S. Chakrabarty and S. P. Kabekkodu, A comprehensive review on the carcinogenic potential of bisphenol A: clues and evidence, *Environ. Sci. Pollut. Res. Int.*, 2021, **28**, 19643–19663.
94. A. Nadal, E. Fuentes, C. Ripoll, S. Villar-Pazos, M. Castellano-Muñoz, S. Soriano, J. Martinez-Pinna, I. Quesada and P. Alonso-Magdalena, Extranuclear-initiated estrogenic actions of endocrine disrupting chemicals: Is there toxicology beyond paracelsus?, *J. Steroid Biochem. Mol. Biol.*, 2018, **176**, 16–22.
95. N. Chevalier, A. Bouskine and P. Fenichel, Bisphenol A promotes testicular seminoma cell proliferation through GPER/GPR30, *Int. J. Cancer*, 2012, **130**, 241–242.
96. S. González-Rojo, M. Lombó, C. Fernández-Díez and M. P. Herráez, Male exposure to bisphenol a impairs spermatogenesis and triggers histone hyperacetylation in zebrafish testes, *Environ. Pollut.*, 2019, **248**, 368–379.
97. Y. Qie, W. Qin, K. Zhao, C. Liu, L. Zhao and L. H. Guo, Environmental Estrogens and Their Biological Effects through GPER Mediated Signal Pathways, *Environ. Pollut.*, 2021, **278**, 116826.
98. S. Xu, S. Yu, D. Dong and L. T. O. Lee, G Protein-Coupled Estrogen Receptor: A Potential Therapeutic Target in Cancer, *Front. Endocrinol.*, 2019, **10**, 725.
99. C. M. Revankar, H. D. Mitchell, A. S. Field, R. Burai, C. Corona, C. Ramesh, L. A. Sklar, J. B. Arterburn and E. R. Prossnitz, Synthetic Estrogen Derivatives Demonstrate the Functionality of Intracellular GPR30, *ACS Chem. Biol.*, 2007, **2**, 536–544.
100. C. Wang, J. Zhang, Q. Li, T. Zhang, Z. Deng, J. Lian, D. Jia, R. Li, T. Zheng, X. Ding, F. Yang, C. Ma, R. Wang, W. Zhang and J. Guo Wen, Low concentration of BPA induces mice spermatocytes apoptosis via GPR30, *Oncotarget*, 2017, **8**, 49005–49015.
101. P. Thomas, Role of G-protein-coupled estrogen receptor (GPER/GPR30) in maintenance of meiotic arrest in fish oocytes, *J. Steroid Biochem. Mol. Biol.*, 2017, **167**, 153–161.
102. S. A. Hafezi and W. M. Abdel-Rahman, The Endocrine Disruptor Bisphenol A (BPA) Exerts a Wide Range of Effects in Carcinogenesis and Response to Therapy, *Curr. Mol. Pharmacol.*, 2019, **12**, 230–238.
103. M. Murata and J. H. Kang, Bisphenol A (BPA) and cell signaling pathways, *Biotechnol. Adv.*, 2018, **36**, 311–327.
104. W. Qu, Z. Zhao, S. Chen, L. Zhang, D. Wu and Z. Chen, Bisphenol A suppresses proliferation and induces apoptosis in colonic epithelial cells through mitochondrial and MAPK/AKT pathways, *Life Sci.*, 2018, **208**, 167–174.
105. S. Dong, S. Terasaka and R. Kiyama, Bisphenol A induces a rapid activation of Erk1/2 through GPR30 in human breast cancer cells, *Environ. Pollut.*, 2011, **159**, 212–218.

106. M. Pupo, A. Pisano, R. Lappano, M. F. Santolla, E. M. De Francesco, S. Abonante, C. Rosano and M. Maggiolini, Bisphenol A induces gene expression changes and proliferative effects through GPER in breast cancer cells and cancer-associated fibroblasts, *Environ. Health Perspect.*, 2012, **120**, 1177–1182.
107. Z. G. Sheng, W. Huang, Y. X. Liu and B. Z. Zhu, Bisphenol A at a low concentration boosts mouse spermatogonial cell proliferation by inducing the G protein-coupled receptor 30 expression, *Toxicol. Appl. Pharmacol.*, 2013, **267**, 88–94.
108. X. L. Zhang, N. Liu, S. F. Weng and H. S. Wang, Bisphenol A Increases the Migration and Invasion of Triple-Negative Breast Cancer Cells via Oestrogen-related Receptor Gamma, *Basic Clin. Pharmacol. Toxicol.*, 2016, **119**, 389–395.
109. I. S. Okon and M. H. Zou, Mitochondrial ROS and cancer drug resistance: Implications for therapy, *Pharmacol. Res.*, 2015, **100**, 170–174.
110. S. J. Sauer, M. Tarpley, I. Shah, A. V. Save, H. K. Lyerly, S. R. Patierno, K. P. Williams and G. R. Devi, Bisphenol A activates EGFR and ERK promoting proliferation, tumor spheroid formation and resistance to EGFR pathway inhibition in estrogen receptor-negative inflammatory breast cancer cells, *Carcinogenesis*, 2017, **38**, 252–260.
111. S. H. Dairkee, M. G. Luciani-Torres, D. H. Moore and W. H. Goodson, 3rd, Bisphenol-A-induced inactivation of the p53 axis underlying deregulation of proliferation kinetics, and cell death in non-malignant human breast epithelial cells, *Carcinogenesis*, 2013, **34**, 703–712.
112. W. H. Goodson 3rd, M. G. Luciani, S. A. Sayeed and I. M. Jaffee, D. H. Moore, 2nd and S. H. Dairkee, Activation of the mTOR pathway by low levels of xenoestrogens in breast epithelial cells from high-risk women, *Carcinogenesis*, 2011, **32**, 1724–1733.
113. K. H. Wang, A. P. Kao, C. C. Chang, T. C. Lin and T. C. Kuo, Bisphenol A-induced epithelial to mesenchymal transition is mediated by cyclooxygenase-2 up-regulation in human endometrial carcinoma cells, *Reprod. Toxicol.*, 2015, **58**, 229–233.
114. S. Li, B. Wang, Q. Tang, J. Liu and X. Yang, Bisphenol A triggers proliferation and migration of laryngeal squamous cell carcinoma via GPER mediated upregulation of IL-6, *Cell Biochem. Funct.*, 2017, **35**, 209–216.
115. K.-S. Zhang, H.-Q. Chen, Y.-S. Chen, K.-F. Qiu, X.-B. Zheng, G.-C. Li, H.-D. Yang and C.-J. Wen, Bisphenol A stimulates human lung cancer cell migration via upregulation of matrix metalloproteinases by GPER/EGFR/ERK1/2 signal pathway, *Biomed. Pharmacother.*, 2014, **68**, 1037–1043.
116. R. Castillo Sanchez, R. Gomez and E. Perez, Salazar, Bisphenol A Induces Migration through a GPER-, FAK-, Src-, and ERK2-Dependent Pathway in MDA-MB-231 Breast Cancer Cells, *Chem. Res. Toxicol.*, 2016, **29**, 285–295.

117. X. F. Ma, J. Zhang, H. L. Shuai, B. Z. Guan, X. Luo and R. L. Yan, IKKβ/NF-κB mediated the low doses of bisphenol A induced migration of cervical cancer cells, *Arch. Biochem. Biophys.*, 2015, **573**, 52–58.
118. A. Ptak and E. L. Gregoraszczuk, Bisphenol A induces leptin receptor expression, creating more binding sites for leptin, and activates the JAK/Stat, MAPK/ERK and PI3K/Akt signalling pathways in human ovarian cancer cell, *Toxicol. Lett.*, 2012, **210**, 332–337.
119. I. Quesada, E. Fuentes, M. C. Viso-León, B. Soria, C. Ripoll and A. Nadal, Low doses of the endocrine disruptor bisphenol-A and the native hormone 17beta-estradiol rapidly activate transcription factor CREB, *FASEB J.*, 2002, **16**, 1671–1673.
120. T. Yaguchi, The endocrine disruptor bisphenol A promotes nuclear ERRγ translocation, facilitating cell proliferation of Grade I endometrial cancer cells via EGF-dependent and EGF-independent pathways, *Mol. Cell. Biochem.*, 2019, **452**, 41–50.
121. M. Huang, M. Huang, X. Li, S. Liu, L. Fu, X. Jiang and M. Yang, Bisphenol A induces apoptosis through GPER-dependent activation of the ROS/Ca(2+)-ASK1-JNK pathway in human granulosa cell line KGN, *Ecotoxicol. Environ. Saf.*, 2021, **208**, 111429.
122. M. García-Arevalo, P. Alonso-Magdalena, J. Rebelo Dos Santos, I. Quesada, E. M. Carneiro and A. Nadal, Exposure to bisphenol-A during pregnancy partially mimics the effects of a high-fat diet altering glucose homeostasis and gene expression in adult male mice, *PLoS One*, 2014, **9**, e100214.
123. M. K. Moon, I. K. Jeong, H. Y. Ahn, H. H. Kim, Y. J. Park, H. C. Jang and K. S. Park, Long-term oral exposure to bisphenol A induces glucose intolerance and insulin resistance, *J. Endocrinol.*, 2015, **226**, 35–42.
124. A. Wang, J. Luo, W. Moore, H. Alkhalidy, L. Wu, J. Zhang, W. Zhen, Y. Wang, D. J. Clegg, X. Bin, Z. Cheng, R. P. McMillan, M. W. Hulver and D. Liu, GPR30 regulates diet-induced adiposity in female mice and adipogenesis in vitro, *Sci. Rep.*, 2016, **6**, 34302.
125. I. Cimmino, F. Oriente, V. D'Esposito, D. Liguoro, P. Liguoro, M. R. Ambrosio, S. Cabaro, F. D'Andrea, F. Beguinot, P. Formisano and R. Valentino, Low-dose Bisphenol-A regulates inflammatory cytokines through GPR30 in mammary adipose cells, *J. Mol. Endocrinol.*, 2019, **63**, 273–283.
126. L. Yu, P. Das, A. J. Vall, Y. Yan, X. Gao, M. I. Sifre, C. D. Bortner, L. Castro, G. E. Kissling, A. B. Moore and D. Dixon, Bisphenol A induces human uterine leiomyoma cell proliferation through membrane-associated ERα36 via nongenomic signaling pathways, *Mol. Cell. Endocrinol.*, 2019, **484**, 59–68.
127. X. Lan, L. J. Fu, J. Zhang, X. Q. Liu, H. J. Zhang, X. Zhang, M. F. Ma, X. M. Chen, J. L. He, L. B. Li, Y. X. Wang and Y. B. Ding, Bisphenol A exposure promotes HTR-8/SVneo cell migration and impairs mouse placentation involving upregulation of integrin-β1 and MMP-9 and

stimulation of MAPK and PI3K signaling pathways, *Oncotarget*, 2017, **8**, 51507–51521.
128. B. Jia, T. Shi, Z. Li, S. Shan, P. Ji and Z. Li, Toxicological effects of bisphenol A exposure-induced cancer cells migration via activating directly integrin β1, *Chemosphere*, 2019, **220**, 783–792.
129. E. Somm, V. M. Schwitzgebel, A. Toulotte, C. R. Cederroth, C. Combescure, S. Nef, M. L. Aubert and P. S. Hüppi, Perinatal exposure to bisphenol a alters early adipogenesis in the rat, *Environ. Health Perspect.*, 2009, **117**, 1549–1555.
130. Y. Pu, J. D. Gingrich, J. P. Steibel and A. Veiga-Lopez, Sex-Specific Modulation of Fetal Adipogenesis by Gestational Bisphenol A and Bisphenol S Exposure, *Endocrinology*, 2017, **158**, 3844–3858.
131. G. Biasiotto, I. Zanella, A. Masserdotti, R. Pedrazzani, M. Papa, L. Caimi and D. Di, Lorenzo, Municipal wastewater affects adipose deposition in male mice and increases 3T3-L1 cell differentiation, *Toxicol. Appl. Pharmacol.*, 2016, **297**, 32–40.
132. F. Ariemma, V. D'Esposito, D. Liguoro, F. Oriente, S. Cabaro, A. Liotti, I. Cimmino, M. Longo, F. Beguinot, P. Formisano and R. Valentino, Low-Dose Bisphenol-A Impairs Adipogenesis and Generates Dysfunctional 3T3-L1 Adipocytes, *PLoS One*, 2016, **11**, e0150762.
133. E. Atlas, L. Pope, M. G. Wade, A. Kawata, A. Boudreau and J. G. Boucher, Bisphenol A increases aP2 expression in 3T3L1 by enhancing the transcriptional activity of nuclear receptors at the promoter, *Adipocyte*, 2014, **3**, 170–179.
134. J. G. Boucher, R. Gagné, A. Rowan-Carroll, A. Boudreau, C. L. Yauk and E. Atlas, Bisphenol A and Bisphenol S Induce Distinct Transcriptional Profiles in Differentiating Human Primary Preadipocytes, *PLoS One*, 2016, **11**, e0163318.
135. J. Wang, B. Sun, M. Hou, X. Pan and X. Li, The environmental obesogen bisphenol A promotes adipogenesis by increasing the amount of 11β-hydroxysteroid dehydrogenase type 1 in the adipose tissue of children, *Int. J. Obes.*, 2013, **37**, 999–1005.
136. A. Salehpour, F. Shidfar, M. Hedayati, A. Neshatbini Tehrani, A. A. Farshad and S. Mohammadi, Bisphenol A enhances adipogenic signaling pathways in human mesenchymal stem cells, *Genes Environ.*, 2020, **42**, 13.
137. B. DeBenedictis, H. Guan and K. Yang, Prenatal Exposure to Bisphenol A Disrupts Mouse Fetal Liver Maturation in a Sex-Specific Manner, *J. Cell. Biochem.*, 2016, **117**, 344–350.
138. Q. Ma, Role of nrf2 in oxidative stress and toxicity, *Annu. Rev. Pharmacol. Toxicol.*, 2013, **53**, 401–426.
139. Y. Dong, Z. Zhang, H. Liu, L. Jia, M. Qin and X. Wang, Exacerbating lupus nephritis following BPA exposure is associated with abnormal autophagy in MRL/lpr mice, *Am. J. Transl. Res.*, 2020, **12**, 649–659.
140. J. Xu, A. C. Donepudi, V. R. More, S. R. Kulkarni, L. Li, L. Guo, B. Yan, T. Chatterjee, N. Weintraub and A. L. Slitt, Deficiency in Nrf2

transcription factor decreases adipose tissue mass and hepatic lipid accumulation in leptin-deficient mice, *Obesity*, 2015, **23**, 335–344.
141. M. Nakamura, H. Yamanaka, A. Oguro and S. Imaoka, Bisphenol A induces Nrf2-dependent drug-metabolizing enzymes through nitrosylation of Keap1, *Drug Metab. Pharmacokinet.*, 2018, **33**, 194–202.
142. P. C. Shimpi, V. R. More, M. Paranjpe, A. C. Donepudi, J. M. Goodrich, D. C. Dolinoy, B. Rubin and A. L. Slitt, Hepatic Lipid Accumulation and Nrf2 Expression following Perinatal and Peripubertal Exposure to Bisphenol A in a Mouse Model of Nonalcoholic Liver Disease, *Environ. Health Perspect.*, 2017, **125**, 087005.
143. Q. Li, J. Davila, A. Kannan, J. A. Flaws, M. K. Bagchi and I. C. Bagchi, Chronic Exposure to Bisphenol A Affects Uterine Function During Early Pregnancy in Mice, *Endocrinology*, 2016, **157**, 1764–1774.
144. Y. L. Schindler, K. M. Garske, J. Wang, B. A. Firulli, A. B. Firulli, K. D. Poss and D. Yelon, Hand2 elevates cardiomyocyte production during zebrafish heart development and regeneration, *Development*, 2014, **141**, 3112–3122.
145. J. Moreman, O. Lee, M. Trznadel, A. David, T. Kudoh and C. R. Tyler, Acute Toxicity, Teratogenic, and Estrogenic Effects of Bisphenol A and Its Alternative Replacements Bisphenol S, Bisphenol F, and Bisphenol AF in Zebrafish Embryo-Larvae, *Environ. Sci. Technol.*, 2017, **51**, 12796–12805.
146. M. Lombó, S. González-Rojo, C. Fernández-Díez and M. P. Herráez, Cardiogenesis impairment promoted by bisphenol A exposure is successfully counteracted by epigallocatechin gallate, *Environ. Pollut.*, 2019, **246**, 1008–1019.
147. M. Featherstone, in *Advances in Developmental Biology and Biochemistry*, Elsevier, 2003, vol. 13, pp. 1–42.
148. T. R. Lappin, D. G. Grier, A. Thompson and H. L. Halliday, HOX genes: seductive science, mysterious mechanisms, *Ulster Med. J.*, 2006, **75**, 23–31.
149. B. Bodey, B. Bodey, Jr., S. E. Siegel and H. E. Kaiser, Immunocytochemical detection of the homeobox B3, B4, and C6 gene products in breast carcinomas, *Anticancer Res.*, 2000, **20**, 3281–3286.
150. T. Hayashida, F. Takahashi, N. Chiba, E. Brachtel, M. Takahashi, N. Godin-Heymann, K. W. Gross, M. d. M. Vivanco, V. Wijendran, T. Shioda, D. Sgroi, P. K. Donahoe and S. Maheswaran, HOXB9, a gene overexpressed in breast cancer, promotes tumorigenicity and lung metastasis, *Proceedings of the National Academy of Sciences*, 2010, **107**, 1100.
151. C. C. Smith and H. S. Taylor, Xenoestrogen exposure imprints expression of genes (Hoxa10) required for normal uterine development, *FASEB J.*, 2007, **21**, 239–246.
152. P. Deb, A. Bhan, I. Hussain, K. I. Ansari, S. A. Bobzean, T. K. Pandita, L. I. Perrotti and S. S. Mandal, Endocrine disrupting chemical, bisphenol-A, induces breast cancer associated gene HOXB9 expression in vitro and in vivo, *Gene*, 2016, **590**, 234–243.

153. I. Hussain, A. Bhan, K. I. Ansari, P. Deb, S. A. Bobzean, L. I. Perrotti and S. S. Mandal, Bisphenol-A induces expression of HOXC6, an estrogen-regulated homeobox-containing gene associated with breast cancer, *Biochim Biophys Acta*, 2015, **1849**, 697–708.
154. J. R. Rochester, Bisphenol A and human health: a review of the literature, *Reprod. Toxicol.*, 2013, **42**, 132–155.
155. H. H. You and G. Song, Review of endocrine disruptors on male and female reproductive systems, *Comp. Biochem. Physiol., Part C: Toxicol. Pharmacol.*, 2021, **244**, 109002.
156. F. S. vom Saal, B. T. Akingbemi, S. M. Belcher, L. S. Birnbaum, D. A. Crain, M. Eriksen, F. Farabollini, L. J. Guillette, Jr., R. Hauser, J. J. Heindel, S. M. Ho, P. A. Hunt, T. Iguchi, S. Jobling, J. Kanno, R. A. Keri, K. E. Knudsen, H. Laufer, G. A. LeBlanc, M. Marcus, J. A. McLachlan, J. P. Myers, A. Nadal, R. R. Newbold, N. Olea, G. S. Prins, C. A. Richter, B. S. Rubin, C. Sonnenschein, A. M. Soto, C. E. Talsness, J. G. Vandenbergh, L. N. Vandenberg, D. R. Walser-Kuntz, C. S. Watson, W. V. Welshons, Y. Wetherill and R. T. Zoeller, Chapel Hill bisphenol A expert panel consensus statement: integration of mechanisms, effects in animals and potential to impact human health at current levels of exposure, *Reprod. Toxicol.*, 2007, **24**, 131–138.
157. S. Singh and S. S.-L. Li, Epigenetic Effects of Environmental Chemicals Bisphenol A and Phthalates, *Int. J. Mol. Sci.*, 2012, **13**, 10143–10153.

CHAPTER 2

Genotoxicity and Mutagenicity of Bisphenol A

SHOEB IKHLAS AND AFIA USMAN*

Department of Molecular Pharmacology, Albert Einstein College of Medicine, NY, USA
*Email: afiausman91@gmail.com

2.1 Introduction

While synthesized initially as a synthetic estrogen, bisphenol A (BPA) has unique properties as a catalyst for polymerization reactions in the plastic industry. It has since been used for the manufacture of polycarbonate plastics or hard plastic and epoxy resins.[1] The utility of BPA in the plastic industry has made it a high production volume chemical that is present in numerous consumer products such as food and drink packaging, toys, plastic bags, dental sealants, medical equipment, and sports safety equipment.[2,3] The massive production of BPA and its leaching from consumer products has led to widespread environmental contamination.[4] This contamination results in chronic human exposure through dietary (food and water) and non-dietary means (inhalation and absorption).[5,6] A study conducted in the United States (US), as part of the National Health and Nutrition Examination Survey, showed BPA detection in 92.6% of human urine samples from a total of 2517 participants, giving a sense of the abundance of human exposure.[7]

In 2008 a report by the US National Toxicology Program raised "some concern for effects on the brain, behavior and prostate gland in fetuses, infants and children".[1] The US Environmental Protection Agency placed BPA

Issues in Toxicology No. 43
Bisphenol A: A Multi-modal Endocrine Disruptor
Edited by Natalie R. Gassman

Published by the Royal Society of Chemistry, www.rsc.org

third-highest in a toxicological priority index among 309 chemicals.[8] These concerns and several BPA toxicological studies resulted in governmental regulations from the US Food and Drug Administration, the European Commission, and the governments of Canada and Japan prohibiting the use of BPA in baby bottles or infant-related products.[4] Since the concerns noted in 2008, numerous studies have been conducted to assess BPA's toxicity and its role as a genotoxicant. Here, we review both *in vitro* and *in vivo* studies on the genotoxic nature of BPA.

Genotoxicity results from genetic alterations, such as DNA strand breakage, chromosomal structure alteration, and changes in the gene expression profile, induced by exogenous and endogenous agents. Genotoxins exert their effects either by directly interacting with DNA molecules and chromosomes or by inducing oxidative stress *via* reactive oxygen species (ROS).[9] Several endpoints are used to evaluate the genotoxic nature of a substance. The most representative tests are the Ames test, mammalian cell gene mutation assay, chromosome aberration assay, sister chromatid exchange, micronuclei assay, single-cell gel electrophoresis (comet assay), and γH2AX foci measurements.[10] *In vitro* and *in vivo* studies on BPA genotoxicity are summarized in Tables 2.1 and 2.2.

2.2 *In Vitro* Studies

BPA genotoxicity has been explored over the past few decades using different assays and methods. Predictive toxicology, using stem cells or their derived tissues, described BPA as a non-genotoxic carcinogen.[11] DNA repair proteins such as 8-oxoguanine glycosylase (OGG1), apurinic/apyrimidinic endonuclease 1 (Ape-1), and MutY Homolog (MyH), which is involved in the base excision repair pathway, were downregulated by BPA in human papillomavirus 18 prostatic cell line RWPE-1 cells.[12]

2.2.1 Ames Test and Mammalian Cell Gene Mutation Assay

The bacterial reverse mutation test or Ames test is an assay used to identify genotoxic mutagens. BPA has not been found to induce mutagenesis in Ames tests carried out in various *Salmonella typhimurium* strains.[13–16] BPA did not induce mutations at concentrations ranging from non-toxic to toxic (Table 2.1).[17] Mutation assays in mammalian systems such as Chinese hamster V79 cells, mouse lymphoma cells, and Syrian hamster embryo cells did not observe any mutations.[17–20]

However, contrary to all the data pointing to the non-mutagenic nature of BPA, one study employing peptide nucleic acid mediated PCR clamping reported BPA-induced mutagenicity in ultraviolet-sensitive human RSa cells derived from human embryo fibroblast cells doubly infected with Simian virus 40 and Rous sarcoma virus at a concentration of 10^{-8} M, which is within the range of its environmental exposure to humans.[21]

Table 2.1 Bisphenol A genotoxicity in *in vitro* studies.[a]

Test system	Treatment doses	Endpoint	Results	Reference
Salmonella typhimurium TA98, TA100, TA1535, TA1537	0.33–333.3 μg per plate	Mutagenicity using Ames test	−	13
Salmonella typhimurium TA97, TA98, TA100, TA102	5–1000 μg per plate	Mutagenicity using Ames test	−	14
Salmonella typhimurium TA1538	0.1–1.0 $\mu g\,\mu L^{-1}$	Mutagenicity using Ames test	−	15
Salmonella typhimurium tester strains TA97a, TA98, TA100, and TA102	500–0 μg per plate	Mutagenicity using Ames test	−	17
Chinese hamster lung fibroblast cells, V79 cells	200–0 μM	Mutagenicity at hprt locus	−	17
L5178Y mouse lymphoma cells	5–50 $\mu g\,mL^{-1}$	Mutagenicity at thymidine kinase locus	−	18
SHE cells	25–200 μM	Mutation at Na^+/K^+	−	19
	50–200 μM		+	
		ATPase locus or the hprt locus		
		Cell transformation		
SHE cells	100 μM	Mutation at Na^+/K^+	−	20
	100 μM	ATPase locus or the hprt locus	+	
		Cell transformation		
Human RSa cells	10^{-2} μM	PNA-mediated PCR clamping	+	21
SHE cells	25–200 μM	Chromosomal aberration	−	19
SHE cells	50–200 μM	Chromosomal aberration	−	20
CHO cells	400 μM	Chromosomal aberration	−	23
CHO-K1 cells	500–600 μM	Chromosomal aberration	+	24
	400–500 μM	Sister chromatid exchange		
MCF-7 cells	0.4–100 $\mu g\,m^{-1}$	Chromosome aberration	+	25

CHO cells	80–120 μM	Chromosome aberration, micronuclei formation	+	26
PNT1a and PC3 cells	200–250 μM	DNA adduct formation	+	40
Male Chinese hamster V79 lung fibroblast cells	200 μM	Micronuclei formation, mitotic arrest	+	43
V79 cells	100–200 μM	Induction of MTOCs, multipolar spindles and multipolar division	+	44
Oocytes from heifers	1 fM–50 μM	Chromosome misalignment and spindle abnormalities	+	45
Hep-2 and MRC-5 cells	4.4 μM and 4.4 nM	Micronuclei formation	+	46
Bovine lymphocytes	10^2 μM	Micronuclei formation	+	47
Human lymphocytes	0.2–0.02 μg mL^{-1}	Micronuclei formation	+	48
AHH-1 cells	12.3–37 μg mL^{-1}	Micronuclei formation	+	49
HepG-2	0.1–10 μmol L^{-1}	DNA damage using comet assay	+	58
184A1, MCF10A, MCF7, and MDA-MB-231 breast cells	10^{-2} μM	γ-H2AX induction	+	59
Human peripheral blood mononuclear cells	0.01–10 μg mL^{-1}	DNA damage using comet assay	+	55
Rat insulinoma INS-1	50–100 μM	DNA damage using comet assay	+	69
Sperm of sterlet	1.75–10 μg L^{-1}	DNA damage using comet assay	+	70
Human liver cancer cell line HepG2	10^{-8}–10^{-6} mol L^{-1}	DNA damage using comet assay	+	51
Human embryonic stem cells line KhES-3 (XY karyotype)	1–0.0625 μM	Support-vector machine which uses qRT-PCR	Non-genotoxic carcinogen	11

[a]CHO: Chinese hamster ovary; MTOC: microtubule-organizing center; PNA: peptide nucleic acid; SHE: Syrian hamster embryo.

Table 2.2 Bisphenol A genotoxicity in *in vivo* studies.

Study animal	Dose	Administration	Endpoint	Results	Reference
CD-1 mice	200 mg kg^{-1}	Intraperitoneal	DNA adduct formation	+	30
CD-1 albino mice	200 mg kg^{-1}	Oral	DNA adduct formation	+	78
Pzh:Sfis mice	5 mg kg^{-1}	Oral	DNA damage using comet assay	+	80
Wistar rats	30 mg kg^{-1}	Subcutaneous	DNA damage using comet assay	+	81
Sprague Dawley rats	200 mg kg^{-1}	Oral	DNA damage using comet assay	+	72
Sprague Dawley Crl:CD rats	0.5 mg kg^{-1}	Oral	γ -H2AX induction	+	73
Holtzman rats	5 mg kg^{-1}	Oral	DNA damage using comet assay	+	82
Chironomus riparius larvae	0.5 mg L^{-1}		DNA damage using comet assay	+	83
Wild-type Oregon-R *Drosophila melanogaster*	0.0025 and 0.005 mg mL^{-1} food media		DNA damage using comet assay	+	86
Holtzman rats	0.01 mg kg^{-1}	Oral	DNA damage using comet assay	+	87

2.2.2 Chromosome Aberration Assay, Sister Chromatid Exchange Assay Aneuploidy Induction, and Cell Transformation

The chromosome aberration test detects structural abnormalities in chromosomes such as breaks and sister chromatid exchange. BPA did not induce chromosome aberrations in Syrian hamster embryo cells.[19,20] However, some studies have found BPA to be positive in chromosome aberration assays, resulting in gaps, breaks, and fragments in Chinese hamster ovary (CHO) cells, clone WBL.[22,23] Along with chromosome aberration, it also induced sister chromatid exchange in CHO-K1 cells.[24] BPA induced aneuploidy and cell transformation in Syrian hamster embryo cells.[19,20] Structural chromosomal abnormalities induced by BPA were reported in MCF-7 breast cancer cells, but not in amniocytes, suggesting that BPA affects cells with altered DNA repair systems.[25] One study using CHO cells found BPA to be negative in the Ames test, but positive in chromosome aberration, micronuclei induction, and DNA strand break induction, suggesting that BPA acts as a clastogen, not a mutagen.[26]

2.2.3 DNA Adduct Formation

DNA adducts are base modifications that can lead to replication or protein synthesis errors causing DNA or protein mutations, promoting carcinogenesis.[27] BPA undergoes metabolism by hydroxylation, and after oxidation, its products can react with DNA to form adducts.[28] An *in vitro* study attributed DNA adduct formation to a metabolite of BPA, bisphenol-*o*-quinone, identifying the same reactive metabolite in an *in vivo* rat study.[29,30] Quinones are reactive agents that can bind to the DNA and form adducts. The BPA metabolite, bisphenol-A 3,4-quinone, was found to react with deoxyguanosine followed by depurination.[31] However, despite these findings, the German MAK commission calculated the DNA binding ability of BPA or its covalent binding index (CBI) to be only 0.01.[30,32] CBI <0.1 indicates that the substance has very low DNA binding ability.[33–36] Additionally, bisphenol-*o*-quinone is not a significant metabolite under physiologically relevant *in vivo* conditions, where BPA mainly undergoes glucuronidation and not oxidation.[37–39] DNA adduct formation has also been reported in cultured human prostatic cell lines, *i.e.* PTN1a and PC3 cells exposed to high-dose BPA.[40]

2.2.4 Micronucleus and Microtubule Assays

Micronucleus assays have been used to determine cytogenetic damage. Micronuclei that appear as extra-nuclear bodies may result from lagging chromosomes or from chromosome fragments due to mitotic errors or DNA damage.[41] BPA resulted in micronuclei induction, disruption of the microtubule complex, mitotic arrest, aberrant spindles, inhibition of microtubule polymerization, spindle abnormalities, chromosome misalignment, and disruption of the mitotic spindle in male Chinese hamster V79 lung fibroblast cells and oocytes.[42–45] BPA induced micronuclei in Hep-2, MRC-5 cells, bovine peripheral lymphocytes, and human lymphocytes[46–48] and modified the functioning of the microtubule-organizing centers of mitotic spindles.[49]

2.2.5 Comet Assay

The comet assay, also called the single-cell gel electrophoresis assay, has been validated as a sensitive and specific assay for quantitation of DNA damage. It evaluates DNA breaks in individual cells. BPA has been reported to increase DNA damage in mouse lymphoma cells, MCF-7, HepG2, MRC-5, and human lymphocytes.[46,50–53] BPA induced DNA damage at non-cytotoxic concentrations in HepG2 cells.[54] BPA was found to cause single-strand breaks in human peripheral blood mononuclear cells.[55]

2.2.6 γH2AX Assay

The γH2AX assay is a new technique compared to the comet assay and micronuclei induction assay. It is based on the detection of histone H2AX phosphorylation. Double-strand breaks result in the induction of histone H2AX phosphorylation into γH2AX. It is an assay that detects genotoxicity at early stages.

In human cells (intestinal cell line: LS174T, hepatoma cell line: HepG2, and renal cell line: ACHN), BPA was non-genotoxic.[56] Contrary to this, BPA was reported to be genotoxic by the induction of γH2AX foci in MCF-7 cells and HepaRG cells.[52,57] At low nanomolar concentration, BPA increased γH2AX in various mammary cells.[58] BPA genotoxicity in mouse embryonic fibroblasts was aggravated by their treatment with cadmium.[59]

Among various types of DNA damage, double-strand breaks are the most cytotoxic and induce genomic instability.[60] BPA induced DNA damage by inducing double-strand breaks.[61] Its inhibitory effect on DNA repair proteins such as DNA-dependent protein kinase may result in its disruptive effect on double-strand repair.[62–64]

2.2.7 Oxidative Stress Induction

One mechanism attributed to the genotoxic nature of BPA is its ability to induce oxidative stress by increasing intracellular ROS.[65,66] ROS are important cell signaling molecules; however, there may be an increase in their presence due to antioxidant defense system changes. ROS oxidize macromolecules, such as RNA, DNA, proteins, and lipids. Electron-rich DNA bases are the targets of ROS for oxidation. This oxidation of DNA results in DNA damage such as single- and double-strand breaks, base modifications, and crosslinking with proteins.[67] This DNA damage can eventually lead to cell death.[68]

BPA has been found to induce ROS-mediated DNA damage in cells such as INS-1, HepG2, and fish spermatozoa.[51,69,70] Repair of oxidized DNA lesions by base exision repair is suppressed by BPA.[71] Antioxidants can act as scavengers of ROS and hence counter their effects. ROS induction by BPA exposure can be countered by antioxidants such as melatonin, silybin, N-acetylcysteine, and pumpkin seed oil, thereby reducing DNA damage.[72–77] Interestingly, BPA can also protect DNA from damage in a cell-free system in the presence of Fe^{2+} ions; contrary to this, it has been shown to have genotoxic effects by inducing ROS in a cellular system, leading to DNA damage.[61]

2.3 *In Vivo* Studies on BPA Genotoxicity

A study reported high levels of DNA adducts in the liver and mammary cells of CD-1 mice exposed to BPA.[78] These DNA adducts can be formed in the reproductive tissues in the organism during its organogenesis and may have lasting effects on the individual.[79] BPA induced DNA strand breaks in lung cells of female mice.[80] Comet assay showed that BPA induced strand breaks in cardiac cells of Wistar rats.[81] Along with the *in vitro* studies reporting its ability to induce DNA damage *via* oxidative stress, BPA has also been found to induce DNA damage in *in vivo* studies.[72,73] Male rats gavaged with BPA showed an increase in sperm DNA damage.[82] DNA damage was also measured using comet assay in the aquatic insect *Chironomus riparius* after BPA exposure.[83] *Daphnia magna* and *Chironomus riparius* were used as potential bioindicator species and were found to develop genotoxicity, as suggested by the comet assay, when exposed to BPA.[84] BPA also reduced hatchability and induced micronuclei in freshwater snails.[85] BPA induced genotoxicity *via* oxidative stress in a *Drosophila melanogaster* model.[86]

BPA exposure caused a significant increase in structural chromosome aberrations in bone marrow cells and DNA damage in blood lymphocytes and micronuclei in polychromatic erythrocytes of rats. Yet a negative Ames test suggested that BPA is a clastogen and not a mutagen.[87] Accidental exposure to BPA due to damaged caging material resulted in meiotic abnormalities such as an increase in aneuploidy in female mice.[88] To investigate this suspected aneugenic activity of BPA, another study was conducted using C57Bl/6 female and (102/ElxC3H/El)F1 male mice trying to replicate the accidental chronic exposure in the previous study; however, they reported no induction of meiotic abnormalities such as hyperploidy and polyploidy.[89] BPA failed to induce chromosomal aberration and micronuclei in bone marrow cells of Swiss albino mice.[90]

Beyond DNA damage and chromosomal aberrations, other genotoxic effects induced by BPA have been observed.[91] BPA induced changes by increasing the expression of the genotoxicity-controlled gene PARP-1 (poly [ADP-ribose] polymerase 1) in exposed rats.[92] PARP-1 protein is involved in DNA repair and, upon activation, results in the post-translational modification of DNA repair proteins. These activities make it an important marker of genotoxicity. In a toxicogenomic analysis, BPA was found to activate lexA, which codes for repressor lexA, recA coding for DNA recombination protein, and other genes involved in DNA damage and repair pathways.[93]

2.4 Conclusion

Currently, there is too much conflicting information about the genotoxicity of BPA to draw clear mechanistic conclusions. Differences in BPA dose, exposure duration, and cell types examined make conclusions difficult to reach. Haighton *et al.* (2002) concluded that BPA did not possess genotoxic activity *in vivo*.[33] The European Food Safety Authority classified the genotoxic effects of BPA to be "unlikely" based on the human and animal evidence available.[94] However, the numerous reports that demonstrate genotoxicity of BPA at low doses cannot be completely discounted and deserve a more thorough study. The genotoxicity of BPA may result from the multiple mechanisms of action it utilizes, making the endpoint and cell-type selection particularly important in determining genotoxicity.[95] In sum, more standardized studies employing more sensitive techniques must be conducted to analyze the genotoxicity of BPA across cell types at physiological doses.

References

1. NTP-CERHR Expert Panel report on the reproductive and developmental toxicity of methanol, *Reprod. Toxicol.* 2004, **18**, 303–390.
2. B. S. Rubin, Bisphenol A: an endocrine disruptor with widespread exposure and multiple effects, *J. Steroid Biochem. Mol. Biol.*, 2011, **127**(1–2), 27–34.
3. T. Geens, L. Goeyens and A. Covaci, Are potential sources for human exposure to bisphenol-A overlooked?, *Int. J. Hyg. Environ. Health*, 2011, **214**, 339–347.
4. A. Usman and M. Ahmad, From BPA to its analogues: Is it a safe journey?, *Chemosphere*, 2016, **158**, 131–142.
5. F. Ariemma, V. D'Esposito, D. Liguoro, F. Oriente, S. Cabaro and A. Liotti, *et al.*, Low-Dose bisphenol-A impairs adipogenesis and generates dysfunctional 3T3-L1 adipocytes, *PLoS One*, 2016, **11**(3), e0150762.
6. T. Geens, D. Aerts, C. Berthot, J. P. Bourguignon, L. Goeyens and P. Lecomte, *et al.*, A review of dietary and non-dietary exposure to bisphenol-A, *Food Chem. Toxicol.*, 2012, **50**, 3725–3740.
7. A. M. Calafat, X. Ye, L.-Y. Wong, J. A. Reidy and L. L. Needham, Exposure of the US population to bisphenol A and 4-tertiary-octylphenol: 2003-2004, *Environ. Health Perspect.*, 2008, **116**(1), 39–44.
8. D. M. Reif, M. T. Martin, S. W. Tan, K. A. Houck, R. S. Judson and A. M. Richard, *et al.*, Endocrine Profiling and Prioritization of Environmental Chemicals Using ToxCast Data, *Environ. Health Perspect.*, 2010, **118**(12), 1714–1720.

9. H. Barabadi, M. Najafi, H. Samadian, A. Azarnezhad, H. Vahidi and M. A. Mahjoub, *et al.*, A systematic review of the genotoxicity and anti-genotoxicity of biologically synthesized metallic nanomaterials: Are green nanoparticles safe enough for clinical marketing?, *Medicina*, 2019, **55**(8), 439.
10. J. García-Lestón Julia, J. Méndez, E. Pásaro and B. Laffon, Genotoxic effects of lead: An updated review, *Environ. Int.*, 2010, **36**, 623–636.
11. J. Yamane, S. Aburatani, S. Imanishi, H. Akanuma, R. Nagano and T. Kato, *et al.*, Prediction of developmental chemical toxicity based on gene networks of human embryonic stem cells, *Nucleic Acids Res.*, 2016, **44**(12), 5515–5528.
12. O. Kose, W. Rachidi, D. Beal, P. Erkekoglu, H. Fayyad-Kazan, Kocer and B. Gumusel, The effects of different bisphenol derivatives on oxidative stress, DNA damage and DNA repair in RWPE-1 cells: A comparative study, *J Appl Toxicol*, 2020, **40**(5), 643–654.
13. S. Haworth, T. Lawlor, K. Mortelmans, W. Speck and E. Zeiger Salmonella mutagenicity test results for 250 chemicals. *Environ Mutagen*, 1983; **5**(1 S): 3–49.
14. J. Takahata, K. Tamakawa, Y. Takahashi, T. Seki, A. Tsunoda, T. Nohmi and T. Sofuni, Mutagenicity of environmental chemicals II, bisphenol A, *Sendai-shi Eisei Kenkyushoho*, 1990, **20**, 245–247.
15. Shell Oil Co. Toxicity Tests with Diphenylol Propane (DPP): Invitro Mutation Studies, 1978, NTIS/OTS 206596, EPA/OTS Doc ID 878214488.
16. JETOC. *Mutagenicity Test Data of Existing Chemical Substances Based On the Toxicity Investigation of The Industrial Safety And Health Law*, (Suppl. 2), 2000 [cited 2021 Jan 13], Available from: https://hero.epa.gov/hero/index.cfm/reference/details/reference_id/628946.
17. H. Schweikl, G. Schmalz and K. Rackebrandt, The mutagenic activity of unpolymerized resin monomers in Salmonella typhimurium and V79 cells, *Mutat. Res.*, 1998, **415**(1–2), 119–130.
18. B. C. Myhr, W. J. Caspary and H. E. Holden, Chemical mutagenesis at the thymidine kinase locus in L5178Y mouse lymphoma cells: Results for 31 coded compounds in the national toxicology program, *Environ. Mol. Mutagen.*, 1991, **18**(1), 51–83.
19. T. Tsutsui, Y. Tamura, E. Yagi, K. Hasegawa, M. Takahashi and N. Maizumi, *et al.*, Bisphenol-A induces cellular transformation, aneuploidy and DNA adduct formation in cultured Syrian hamster embryo cells, *Int. J. Cancer*, 1998, **75**(2), 290–294.
20. T. Tsutsui, Y. Tamura, A. Suzuki, Y. Hirose, M. Kobayashi and H. Nishimura, *et al.*, Mammalian cell transformation and aneuploidy induced by five bisphenols, *Int. J. Cancer*, 2000, **86**(2), 151–154.
21. S. Takahashi, X. J. Chi, Y. Yamaguchi, H. Suzuki, S. Sugaya and K. Kita, *et al.*, Mutagenicity of bisphenol A and its suppression by interferon-α in human RSa cells, *Mutat. Res., Genet. Toxicol. Environ. Mutagen.*, 2001, **490**(2), 199–207.

22. C. A. Hilliard, M. J. Armstrong, C. I. Bradt, R. B. Hill, S. K. Greenwood and S. M. Galloway, Chromosome aberrations in vitro related to cytotoxicity of non-mutagenic chemicals and metabolic poisons, *Environ. Mol. Mutagen.*, 1998, **31**(4), 316–326.
23. S. M. Galloway, J. E. Miller, M. J. Armstrong, C. L. Bean, T. R. Skopek and W. W. Nichols, DNA synthesis inhibition as an indirect mechanism of chromosome aberrations: Comparison of DNA-reactive and non-DNA-reactive clastogens, *Mutat. Res., Fundam. Mol. Mech. Mutagen.*, 1998, **400**(1–2), 169–186.
24. S. Tayama, Y. Nakagawa and K. Tayama, Genotoxic effects of environmental estrogen-like compounds in CHO-K1 cells, *Mutat. Res., Genet. Toxicol. Environ. Mutagen.*, 2008, **649**(1–2), 114–125.
25. S. M. Aghajanpour-Mir, E. Zabihi, H. Akhavan-Niaki, E. Keyhani, I. Bagherizadeh and S. Biglari, *et al.*, The Genotoxic and Cytotoxic Effects of Bisphenol-A (BPA) in MCF-7 Cell Line and Amniocytes [Internet], *Int. J. Mol. Cell. Med.*, 2016, 19–29.
26. L. Xin, Y. Lin, A. Wang, W. Zhu, Y. Liang and X. Su, *et al.*, Cytogenetic evaluation for the genotoxicity of bisphenol-A in Chinese hamster ovary cells, *Environ. Toxicol. Pharmacol.*, 2015, **40**(2), 524–529.
27. A. K. Basu, DNA damage, mutagenesis and cancer, *Int. J. Mol. Sci.*, 2018, **19**, 970.
28. E. L. Cavalieri and E. G. Rogan, Is bisphenol A a weak carcinogen like the natural estrogens and diethylstilbestrol?, *IUBMB Life*, 2010, **62**(10), 746–751.
29. A. Atkinson and D. Roy, In vitro conversion of environmental estrogenic chemical bisphenol A to DNA binding metabolite(s), *Biochem. Biophys. Res. Commun.*, 1995, **210**(2), 424–433.
30. A. Atkinson and D. Roy, In vivo DNA adduct formation by bisphenol A, *Environ. Mol. Mutagen.*, 1995, **26**(1), 60–66.
31. K. Kolšek, J. Mavri and M. S. Dolenc, Reactivity of bisphenol A-3,4-quinone with DNA. A quantum chemical study, *Toxicol. In Vitro*, 2012, **26**(1), 102–106.
32. G. M. Comission, *Critical Data Evaluation forMAK Values and Classification of Carcinogens*, Wiley-VCH, Weinhein/New York/Chichester/Brisbane/Singapore/Toronto, 1996, 13.
33. L. A. Haighton, J. J. Hlywka, J. Doull, R. Kroes, B. S. Lynch and I. C. Munro, An evaluation of the possible carcinogenicity of bisphenol A to humans, *Regul. Toxicol. Pharmacol.*, 2002, **35**(2 I), 238–254.
34. W. K. Lutz, In vivo covalent binding of organic chemicals to DNA as a quantitative indicator in the process of chemical carcinogenesis, *Mutat. Res., Genet. Toxicol. Environ. Mutagen.*, 1979, **65**, 289–356.
35. W. K. Lutz, Structural characteristics of compounds that can be activated to chemically reactive metabolites: Use for a prediction of a carcinogenic potential, *Arch. Toxicol.*, 1984, **55**(Suppl. 7), 194–207.
36. W. K. Lutz, Quantitative evaluation of DNA binding data for risk estimation and for classification of direct and indirect carcinogens, *J. Cancer Res. Clin. Oncol.*, 1986, **112**(2), 85–91.

37. J. B. Knaak and L. J. Sullivan, Metabolism of bisphenol A in the rat, *Toxicol. Appl. Pharmacol.*, 1966 Mar 1, **8**(2), 175–184.
38. L. H. Pottenger, The Relative Bioavailability and Metabolism of Bisphenol A in Rats Is Dependent upon the Route of Administration, *Toxicol. Sci.*, 2000, **54**(1), 3–18.
39. Y. Nakagawa and S. Tayama, Metabolism and cytotoxicity of bisphenol A and other bisphenols in isolated rat hepatocytes, *Arch. Toxicol.*, 2000, **74**(2), 99–105.
40. S. De Flora, R. T. Micale, S. La Maestra, A. Izzotti, F. D'Agostini and A. Camoirano, *et al.*, Upregulation of Clusterin in Prostate and DNA Damage in Spermatozoa from Bisphenol A–Treated Rats and Formation of DNA Adducts in Cultured Human Prostatic Cells, *Toxicol. Sci*, 2011, **122**(1), 45–51.
41. M. Kwon, M. L. Leibowitz and J. H. Lee, Small but mighty: the causes and consequences of micronucleus rupture, *Exp. Mol. Med.*, 2020, **52**, 1777–1786.
42. M. Metzler and E. Pfeiffer, Effects of estrogens on microtubule polymerization in vitro: Correlation with estrogenicity, *Environ. Health Perspect.*, 1995, 21–22.
43. E. Pfeiffer, B. Rosenberg, S. Deuschel and M. Metzler, Interference with microtubules and induction of micronuclei in vitro by various bisphenols, *Mutat. Res.*, 1997, **390**(1–2), 21–31.
44. T. Ochi, Induction of multiple microtubule-organizing centers, multipolar spindles and multipolar division in cultured V79 cells exposed to diethylstilbestrol, estradiol-17β and bisphenol A, *Mutat. Res., Fundam. Mol. Mech. Mutagen.*, 1999, **431**(1), 105–121.
45. K. A. Campen, K. M. Kucharczyk, B. Bogin, J. M. Ehrlich and C. M. H. Combelles, Spindle abnormalities and chromosome misalignment in bovine oocytes after exposure to low doses of bisphenol A or bisphenol S, *Hum. Reprod.*, 2018, **33**(5), 895–904.
46. C. Ramos, C. Ladeira, S. Zeferino, A. Dias, I. Faria and E. Cristovam, *et al.*, Cytotoxic and genotoxic effects of environmental relevant concentrations of bisphenol A and interactions with doxorubicin, *Mutat. Res., Genet. Toxicol. Environ. Mutagen.*, 2019, **838**, 28–36.
47. I. Šutiaková, N. Kovalkovičová and V. Šutiak, Micronucleus assay in bovine lymphocytes after exposure to bisphenol A in vitro, *In Vitro Cell. Dev. Biol.: Anim.*, 2014, **50**(6), 502–506.
48. A. Santovito, E. Cannarsa, D. Schleicherova and P. Cervella, Clastogenic effects of bisphenol A on human cultured lymphocytes, *Hum. Exp. Toxicol.*, 2018, **37**(1), 69–77.
49. G. E. Johnson and E. M. Parry, Mechanistic investigations of low dose exposures to the genotoxic compounds bisphenol-A and rotenone, *Mutat. Res., Genet. Toxicol. Environ. Mutagen.*, 2008, **651**(1–2), 56–63.
50. M. Lee, J. Kwon and M.-K. Chung, Enhanced prediction of potential rodent carcinogenicity by utilizing comet assay and apoptotic assay in combination, *Mutat. Res.*, 2003, **541**(1–2), 9–19.

51. X. Li, P. Yin and L. Zhao, Effects of individual and combined toxicity of bisphenol A, dibutyl phthalate and cadmium on oxidative stress and genotoxicity in HepG 2 cells, *Food Chem. Toxicol.*, 2017, **105**, 73–81.
52. T. Iso, T. Watanabe, T. Iwamoto, A. Shimamoto and Y. Furuichi, DNA damage caused by bisphenol A and estradiol through estrogenic activity, *Biol. Pharm. Bull.*, 2006, **29**(2), 206–210.
53. B. Lei, J. Xu, W. Peng, Y. Wen, X. Zeng and Z. Yu, *et al.*, In vitro profiling of toxicity and endocrine disrupting effects of bisphenol analogues by employing MCF-7 cells and two-hybrid yeast bioassay, *Environ. Toxicol.*, 2017, **32**(1), 278–289.
54. A. Fic, B. Žegura, M. Sollner Dolenc, M. Filipič and L. Peterlin Mašič, Mutagenicity and DNA damage of bisphenol A and its structural analogues in HepG2 cells, *Arh Hig Rada Toksikol*, 2013, **64**(2), 189–200.
55. K. Mokra, A. Kuźmińska-Surowaniec, K. Woźniak and J. Michałowicz, Evaluation of DNA-damaging potential of bisphenol A and its selected analogs in human peripheral blood mononuclear cells (in vitro study), *Food Chem. Toxicol.*, 2017, **100**, 62–69.
56. M. Audebert, L. Dolo, E. Perdu, J.-P. Cravedi and D. Zalko, Use of the γH2AX assay for assessing the genotoxicity of bisphenol A and bisphenol F in human cell lines, *Arch Toxicol.*, 2011, **85**(11), 1463–1473.
57. N. Quesnot, K. Rondel, M. Audebert, S. Martinais, D. Glaise and F. Morel, *et al.*, Evaluation of genotoxicity using automated detection of γH2AX in metabolically competent HepaRG cells, *Mutagenesis*, 2016, **31**(1), 43–50.
58. D. Pfeifer, Y. M. Chung and M. C. T. Hu, Effects of low-dose bisphenol A on DNA damage and proliferation of breast cells: The role of c-Myc, *Environ. Health Perspect.*, 2015, **123**(12), 1271–1279.
59. Z. Y. Chen, C. Liu, Y. H. Lu, L. L. Yang, M. Li and M. D. He, *et al.*, Cadmium Exposure Enhances Bisphenol A-Induced Genotoxicity through 8-Oxoguanine-DNA Glycosylase-1 OGG1 Inhibition in NIH3T3 Fibroblast Cells, *Cell Physiol. Biochem.*, 2016, **39**(3), 961–974.
60. K. Rothkamm and M. Löbrich, Evidence for a lack of DNA double-strand break repair in human cells exposed to very low x-ray doses, *Proc. Natl. Acad. Sci. U. S. A.*, 2003, **100**(9), 5057–5062.
61. I. Durovcova, J. Spackova, M. Puskar, E. Galova and A. Sevcovicova, Bisphenol a as an environmental pollutant with dual genotoxic and DNA-protective effects, *Neuroendocrinol. Lett.*, 2018, **39**(4), 294–298.
62. P. Allard and M. P. Colaiácovo, Bisphenol A impairs the double-strand break repair machinery in the germline and causes chromosome abnormalities, *Proc. Natl. Acad. Sci. U. S. A.*, 2010, **107**(47), 20405–20410.
63. Y. Ito, T. Ito, S. Karasawa, T. Enomoto, A. Nashimoto and Y. Hase, *et al.*, Identification of DNA-Dependent Protein Kinase Catalytic Subunit (DNA-PKcs) as a Novel Target of Bisphenol A, *PLoS One*, 2012, 7(12), e50481.
64. N. R. Gassman, E. Coskun, P. Jaruga, M. Dizdaroglu and S. H. Wilson, Combined effects of high-dose bisphenol A and oxidizing agent (KBrO3)

on cellular microenvironment, gene expression, and chromatin structure of Ku70-deficient mouse embryonic fibroblasts, *Environ. Health Perspect.*, 2016, **124**(8), 1241–1252.
65. N. R. Gassman, Induction of oxidative stress by bisphenol A and its pleiotropic effects, *Environ. Mol. Mutagen.*, 2017, 60–71.
66. P. Kovacic, How safe is bisphenol A? Fundamentals of toxicity: Metabolism, electron transfer and oxidative stress, *Med. Hypotheses*, 2010, **75**, 1–4.
67. T. W. Yu and D. Anderson, Reactive oxygen species-induced DNA damage and its modification: A chemical investigation, *Mutat. Res., Fundam. Mol. Mech. Mutagen.*, 1997, **379**(2), 201–210.
68. O. Surova and B. Zhivotovsky, Various modes of cell death induced by DNA damage, *Oncogene*, 2013, **32**, 3789–3797.
69. F. Xin, L. Jiang, X. Liu, C. Geng, W. Wang and L. Zhong, *et al.*, Bisphenol A induces oxidative stress-associated DNA damage in INS-1 cells, *Mutat. Res., Genet. Toxicol. Environ. Mutagen.*, 2014, **769**, 29–33.
70. M. Hulak, I. Gazo, A. Shaliutina and P. Linhartova, In vitro effects of bisphenol A on the quality parameters, oxidative stress, DNA integrity and adenosine triphosphate content in sterlet (Acipenser ruthenus) spermatozoa, *Comp. Biochem. Physiol., Part C: Toxicol. Pharmacol.*, 2013, **158**(2), 64–71.
71. N. R. Gassman, E. Coskun, D. F. Stefanick, J. K. Horton, P. Jaruga and M. Dizdaroglu, *et al.*, Bisphenol a promotes cell survival following oxidative DNA damage in mouse fibroblasts, *PLoS One*, 2015, **10**(2), e0118819.
72. H. J. Wu, C. Liu, W. X. Duan, S. C. Xu, M. D. He and C. H. Chen, *et al.*, Melatonin ameliorates bisphenol A-induced DNA damage in the germ cells of adult male rats, *Mutat. Res., Genet. Toxicol. Environ. Mutagen.*, 2013, **752**(1–2), 57–67.
73. S. Kim, E. Choi, M. Kim, J. S. Jeong, K. W. Kang, S. Jee, K. M. Lim and Y. S. Lee, Submicromolar bisphenol A induces proliferation and DNA damage in human hepatocyte cell lines in vitro and in juvenile rats in vivo, *Food Chem. Toxicol.*, 2018, **111**, 125–132.
74. M. Zhang, X. Dai, Y. Lu, Y. Miao, C. Zhou and Z. Cui, *et al.*, Melatonin protects oocyte quality from Bisphenol A-induced deterioration in the mouse, *J. Pineal Res.*, 2017, **62**(3), e12396.
75. F. M. Huang, Y. C. Chang, S. S. Lee, Y. C. Ho, M. L. Yang and H. W. Lin, *et al.*, Bisphenol A exhibits cytotoxic or genotoxic potential via oxidative stress-associated mitochondrial apoptotic pathway in murine macrophages, *Food Chem. Toxicol.*, 2018, **122**, 215–224.
76. S. Lama, D. Vanacore, N. Diano, C. Nicolucci, S. Errico and M. Dallio, *et al.*, Ameliorative effect of Silybin on bisphenol A induced oxidative stress, cell proliferation and steroid hormones oxidation in HepG2 cell cultures, *Sci. Rep.*, 2019, **9**(1), 1–10.
77. E. I. Fawzy, A. I. El Makawy, M. M. El-Bamby and H. O. Elhamalawy, Improved effect of pumpkin seed oil against the bisphenol-A adverse effects in male mice, *Toxicol. Rep.*, 2018, **5**, 857–863.

78. A. Izzotti, S. Kanitz, F. D'Agostini, A. Camoirano and S. De Flora, Formation of adducts by bisphenol A, an endocrine disruptor, in DNA in vitro and in liver and mammary tissue of mice, *Mutat. Res., Genet. Toxicol. Environ. Mutagen.*, 2009, **679**(1–2), 28–32.
79. N. Ben-Jonathan and R. Steinmetz, Xenoestrogens: the emerging story of bisphenol a, *Trends Endocrinol Metab.*, 1998, **9**(3), 124–128.
80. A. Gajowik, J. Radzikowska and M. M. Dobrzyńska, Genotoxic effects of bisphenol A on somatic cells of female mice, alone and in combination with X-rays, *Mutat. Res., Genet. Toxicol. Environ. Mutagen.*, 2013, **757**(2), 120–124.
81. D. M. Amin, Role of copeptin as a novel biomarker of bisphenol A toxic effects on cardiac tissues: biochemical, histological, immunohistological, and genotoxic study, *Environ. Sci. Pollut. Res.*, 2019, **26**(35), 36037–36047.
82. D. Tiwari and G. Vanage, Mutagenic effect of Bisphenol A on adult rat male germ cells and their fertility, *Reprod. Toxicol.*, 2013, **40**, 60–68.
83. P. Martínez-Paz, M. Morales, J. L. Martínez-Guitarte and G. Morcillo, Genotoxic effects of environmental endocrine disruptors on the aquatic insect Chironomus riparius evaluated using the comet assay, *Mutat. Res., Genet. Toxicol. Environ. Mutagen.*, 2013, **758**(1–2), 41–47.
84. S. Y. Park and J. Choi, Genotoxic effects of nonylphenol and bisphenol a exposure in aquatic biomonitoring species: Freshwater crustacean, daphnia magna, and aquatic midge, chironomus riparius, *Bull. Environ. Contam. Toxicol.*, 2009, **83**(4), 463–468.
85. P. Sánchez-Argüello, N. Aparicio and C. Fernández, Linking embryo toxicity with genotoxic responses in the freshwater snail Physa acuta: Single exposure to benzo(a)pyrene, fluoxetine, bisphenol A, vinclozolin and exposure to binary mixtures with benzo(a)pyrene, *Ecotoxicol. Environ. Saf.*, 2012, **80**, 152–160.
86. A. Anet, S. Olakkaran, A. Kizhakke Purayil and G. Hunasanahally Puttaswamygowda, Bisphenol A induced oxidative stress mediated genotoxicity in Drosophila melanogaster, *J. Hazard. Mater.*, 2019, **370**, 42–53.
87. D. Tiwari, J. Kamble, S. Chilgunde, P. Patil, G. Maru and D. Kawle, *et al.*, Clastogenic and mutagenic effects of bisphenol A: An endocrine disruptor, *Mutat. Res., Genet. Toxicol. Environ. Mutagen.*, 2012, **743**(1–2), 83–90.
88. P. A. Hunt, K. E. Koehler, M. Susiarjo, C. A. Hodges, A. Ilagan and R. C. Voigt, *et al.*, Bisphenol a exposure causes meiotic aneuploidy in the female mouse, *Curr. Biol.*, 2003, **13**(7), 546–553.
89. F. Pacchierotti, R. Ranaldi, U. Eichenlaub-Ritter, S. Attia and I.-D. Adler, Evaluation of aneugenic effects of bisphenol A in somatic and germ cells of the mouse, *Mutat. Res.*, 2008, **651**(1–2), 64–70.
90. P. Naik and K. K. K. Vijayalaxmi, Cytogenetic evaluation for genotoxicity of bisphenol-A in bone marrow cells of Swiss albino mice, *Mutat. Res.*, 2009, **676**(1–2), 106–112.

91. O. K. Ulutaş, N. Yildiz, E. Durmaz, M. A. Ahbab, N. Barlas and İ. Çok, An in vivo assessment of the genotoxic potential of bisphenol A and 4-tert-octylphenol in rats, *Arch. Toxicol.*, 2011, **85**(8), 995–1001.
92. M. Novo, I. Verdú, D. Trigo and J. L. Martínez-Guitarte, Endocrine disruptors in soil: Effects of bisphenol A on gene expression of the earthworm Eisenia fetida, *Ecotoxicol. Environ. Saf.*, 2018, **150**, 159–167.
93. N. Gou, S. Yuan, J. Lan, C. Gao, A. N. Alshawabkeh and A. Z. Gu, A Quantitative toxicogenomics assay reveals the evolution and nature of toxicity during the transformation of environmental pollutants, *Environ. Sci. Technol.*, 2014, **48**(15), 8855–8863.
94. Authority E food and safety, Draft Bisphenol A (BPA) hazard assessment protocol, 2017.
95. E. Ribeiro, C. Ladeira and S. Viegas, Occupational exposure to Bisphenol A (BPA): A reality that still needs to be unveiled, *Toxics*, 2017, **5**(3), 22.

CHAPTER 3

Bisphenol A Associated Signaling Pathways in Human Diseases

DIVYA ADIGA,[a] G. NADEEM KHAN,[a] SANGAVI ESWARAN,[a] S. SRIHARIKRISHNAA,[a] SANJIBAN CHAKRABARTY,[a,c] PADMALATHA S. RAI[b] AND SHAMA PRASADA KABEKKODU*[a,c]

[a] Department of Cell and Molecular Biology, Manipal School of Life Sciences, Manipal Academy of Higher Education, Manipal, Karnataka, India; [b] Department of Biotechnology, Manipal School of Life Sciences, Manipal Academy of Higher Education, Manipal, Karnataka, India; [c] Center for DNA Repair and Genome Stability (CDRGS), Manipal Academy of Higher Education, Manipal, India
*Email: shama.prasada@manipal.edu

3.1 Introduction

A multitude of synthetic chemicals are being released into the environment, most of which are considered to have potential adverse effects on human health. Bisphenol A (BPA) is one such chemical contaminant found ubiquitously in terrestrial and aquatic environments.[1,2] It is a solid, organic, colorless compound with the chemical formula $(CH_3)_2C(C_6H_4OH)_2$ (molecular weight 228.29 $g\,cm^{-3}$). BPA is a chemical strengthener used in the manufacture of polycarbonate plastics (PCP) (~65% of global BPA demand), epoxy resins (~28%), and flame retardants (~7%).[3] It was first synthesized by a Russian chemist, Alexander Dianin in 1891.[4] The global

Issues in Toxicology No. 43
Bisphenol A: A Multi-modal Endocrine Disruptor
Edited by Natalie R. Gassman

Published by the Royal Society of Chemistry, www.rsc.org

production volume of BPA was assessed at 7.7 million metric tons in 2015 and is projected to reach 10.6 million metric tons by 2022. China, India, Russia, and Mexico are the major producers of BPA, possibly due to the extensive use of PCP products and epoxy resins.[2] PCPs are used to manufacture water bottles, feeding bottles, toys, water pipes, food cans, thermal paper, and medical equipment. Due to their thermal stability, epoxy resins are mostly used in food and beverage containers as internal protective coatings, paints, electronic and electrical materials, adhesives, and in packaging.[5,6] Through all these means, more than a billion pounds of BPA is released into the environment every year.

In the 1930s, estrogen-mimicking characteristics of BPA were determined, but were not given considerable attention until the discovery of nuclear hormone receptors.[7] The evolving concern over environmental estrogens and the discovery of estrogenic BPA leachates introduced BPA to the emerging field of endocrine disruption. Endocrine-disrupting chemicals are effective at low doses and can interrupt or mimic the action of endogenous hormones. The distinctive properties of BPA do not support the theory of "the dose makes the poison".[8,9] Its low-dose effects, rapid metabolism, and nonmonotonic dose response curves defy the conventional principles of toxicology.[10] BPA is still being used worldwide despite the longstanding concern over its safety, as it has been challenging to develop safe and economical replacements. Currently, it is one of the most abundant industrial synthetic chemicals produced globally ($>6 \times 10^9$ lb per year).[11]

3.2 Sources of Human Exposure and Metabolism of BPA

Soil, air, and the aquatic environment can be the source of BPA exposure to various forms of life. Food and drinking water are believed to be the main source of exposure to humans. BPA can leach out from food and beverage containers under various condition.[1,12] A study has reported the presence of 0.234 ng mL^{-1} of BPA in water stored in PCP bottles at room temperature for 5 days, and the amount of leachate can elevate with the rise in storage temperature.[13] Similarly, heating or sterilizing at a high temperature can facilitate BPA leach out of food cans. Elsewhere, dermal exposure to BPA was suggested through thermal printed paper.[6,14] In rodent models and humans, gestational and lactational exposures are evidenced at or below the lowest observed adverse effect level of 0.05 mg kg^{-1} body weight (BW) per day.[15] The probable daily tolerable intake of BPA ranges from <1 to 5 μg kg^{-1} BW per day.[16] Upon absorption, BPA is conjugated by glucuronidation or sulfation in the liver and gets eliminated through feces, bile, and urine and has an average half-life of 6 h in humans. BPA can accumulate in the adipose tissue because of its lipophilic nature. Gender, physiological conditions, age, and liver functional status can impact its metabolism.[17,18]

Owing to its structural similarity to estrogen, BPA can compete with estrogen in binding to the estrogen receptors α and β (ERα and ERβ, respectively) and hence can modulate the estrogen-responsive signaling cascade.[7] In addition to ER receptors, BPA can also bind to estrogen-related receptors (ERR), androgen receptor, pregnane X receptor, thyroid hormone receptor, peroxisome proliferator-activated receptors (PPARs), and aryl hydrocarbon receptors that are associated with the endocrine system.[19] Many experimental and epidemiological studies have described the association between high free BPA levels in body fluids with different chronic health complications such as obesity, cancer, cardiovascular disorders, hepatotoxicity, diabetes, neurological, reproductive, and developmental disorders[12] (Figure 3.1). Several signaling cascades are implicated in the toxic or endocrine-disruptive effects of BPA.[20] This chapter describes the molecular alterations associated with BPA exposure and different human ailments, with a detailed note on the intricate signaling cascades involved in BPA-mediated carcinogenesis. The various molecular mechanisms linking BPA and human diseases are described.

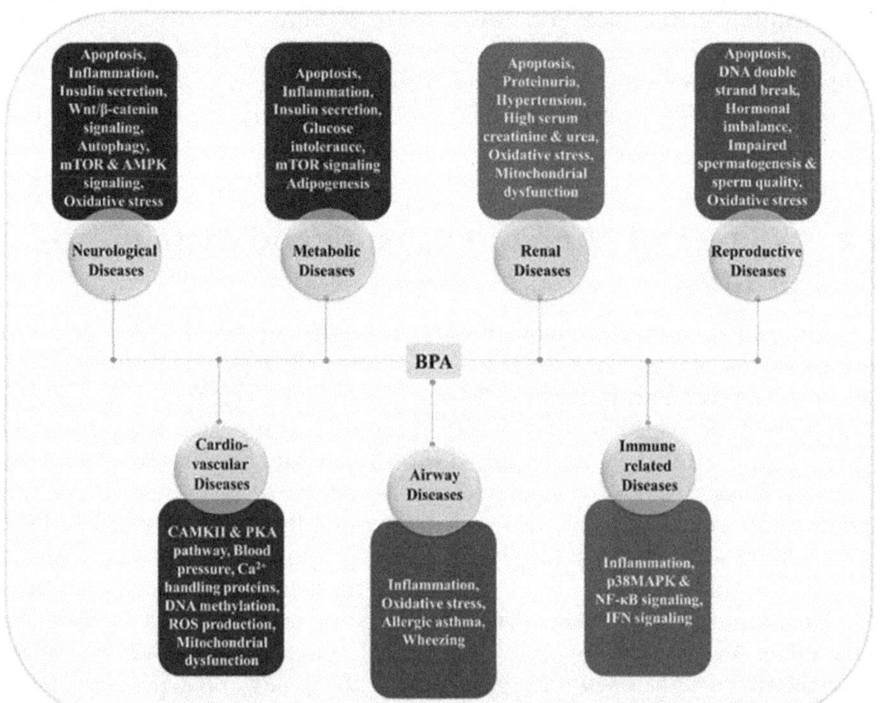

Figure 3.1 BPA-associated human health complications. BPA exposure can increase the risk of several chronic health complications such as cardiovascular, respiratory, neurological, reproductive and metabolic disorders by interfering with the intricate signaling pathways.

3.3 Mechanisms Linking BPA and Human Diseases

Elevated BPA exposure has been linked to a number of human ailments.[6,12] BPA-induced toxic effects, molecular mechanisms, and disease modeling were investigated using various *in vitro* and *in vivo* models. BPA can exert its toxic effects by inducing genetic and epigenetic alterations, oxidative stress, mitochondrial disfunction, endocrine disruption, and by modulating cellular signaling pathways[21] (Figure 3.2). Experimental and epidemiological study models have reported the ability of BPA to induce mutagenicity, aneuploidy, and DNA adducts, which in turn can promote miscarriage, infertility, and birth defects in humans.[21,22] In 2010, Allard and colleagues showed that BPA exposure can result in compromised chromosome synapsis, downregulation of meiotic double-strand break repair genes, and hence impaired genome integrity at meiosis.[23] BPA-mediated alteration in DNA methylation was first demonstrated in 2006 when neonatal BPA exposure enhanced rat susceptibility to neoplastic lesions in prostate tissues by inducing hypomethylation of the phosphodiesterase type 4 variant 4 (*PDE4D4*) gene and increased expression.[24] Reduced methylation of Homeobox A10 (*HOXA10*) was reported in BPA-treated CD-1 mice.[25] Downregulation of the lysosomal-associated membrane protein 3 (*LAMP3*) gene by increased DNA methylation at CpG islands was detected upon low-dose BPA exposure in human primary breast epithelial cells, indicating the risk of breast cancer development later in life.[26] In developing hypothalamic cells, BPA was demonstrated to interfere with the level of methyl CpG binding protein 2 (MECP2) and DNA methyl transferases in a dose-dependent manner.[27] Prenatal exposure of mammary gland cells to BPA elevates trimethylation of histone 3 (H3) at lysine 27 (H3K27me3) and augments the level of histone methyltransferase enhancer of zeste homolog 2 (EZH2).[28] Furthermore, BPA was shown to induce trimethylation of histone H3K4 at the transcriptional start site of the alpha-lactalbumin (LALBA) gene, resulting in its hyperactivation.[29] Studies have depicted that BPA can potentially impact developmental pathways and the reproductive system by altering miRNA expression. In human placental cell lines (miR-146a) and fetal ewe ovaries, altered miRNA expression was reported upon BPA exposure.[30] Prenatal exposure to BPA affects insulin homeostasis in sheep fetal ovaries by modulating miRNAs involved in insulin signaling regulation.[31]

Diminished antioxidant defense and/or elevated reactive oxygen species (ROS) production can disturb the oxidative balance, which can modify the cellular components to activate or inactivate the signaling pathways.[32] Growing evidence indicates that BPA can disrupt the oxidative balance of the cell, leading to cellular damage and oxidative stress. Studies have revealed that BPA can disrupt the oxidative balance by modulating antioxidant enzymes (superoxide dismutase, catalase, glutathione (GSH) peroxidase), mitochondrial functions, and enhancing thiobarbituric acid-reactive substances in mouse brain, kidney, and testes.[21] BPA exposure was shown to stimulate oxidative stress and inflammation in both male and female reproductive organs. A positive correlation between urinary BPA and C-reactive protein (CRP),

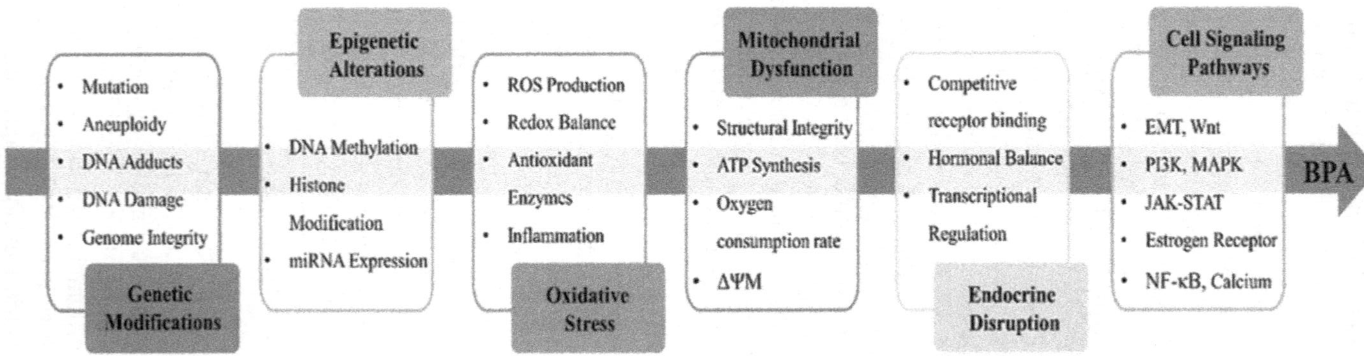

Figure 3.2 Schematic representation of different mechanisms that correlate BPA and human diseases. BPA has been shown to be associated with several human diseases by inducing genetic and epigenetic alterations, oxidative stress, mitochondrial dysfunction, and endocrine disruption, and by regulating cellular signaling pathways.

8-hydroxydeoxyguanosine (8-OHdG), and malondialdehyde (MDA) levels has been described.[33] BPA-mediated oxidative stress was reported to hasten human islet amyloid polypeptide (hIAPP) aggregation and leads to pancreatic β-cell death.[34] Additionally, BPA exposure can impair mitochondrial function by altering mitochondrial membrane potential (ΔΨM), mitochondrial mass, adenosine triphosphate (ATP) depletion, cytochrome c release, and by modifying mitochondrial metabolism and dynamics-associated gene expression.[35]

Endocrine disruption is a mechanism that can be implicated at distinct physiological levels by changing hormone secretion, and/or by altering circulating hormone metabolism, and/or by interrupting receptor–hormone interaction.[36] Interference in the endocrine balance can cause a variety of adverse effects such as reproductive and developmental disorders, hormone-dependent cancers, birth defects, metabolic, neurological, and immune complications.[12,36] BPA's ability to bind to nuclear ERs, classical and non-classical membrane ERs, G-protein-coupled receptor 30 (GPR30), and function *via* non-genomic pathways classify it as a xenoestrogen.[37] With the ability to bind to different receptors, BPA can bring in various changes in the target organs of estrogen such as the mammary gland, testis, ovary, uterus, and brain.[12] Besides, BPA can inhibit ERβ ubiquitination and degradation, thereby affecting ERβ-mediated target gene transcription.[38] Furthermore, BPA can interfere with glucocorticoid receptors, androgen receptors, ERRγ, the dopaminergic system, and thyroid hormone pathways, resulting in behavioral disorders. BPA can even affect cellular signaling cascades by regulating cell Ca^{2+} levels and activation of different signal transducing molecules such as serine/threonine protein kinase (AKT), extracellular signal-regulated kinase (ERK), c-Jun N-terminal kinase (JNK), and nuclear factor-κ light chain enhancer of activated B cells (NF-κB).[20,39,40] At very low concentrations, BPA can affect hormonal mechanisms and their downstream functional responses by triggering several kinases.[41] Details of BPA-mediated regulation of cellular signaling are given in the subsequent sections.

3.4 BPA and Human Diseases

Numerous epidemiological studies have demonstrated the positive correlation between BPA levels and human diseases. The National Toxicology Program, the National Center for Toxicological Research of the United States Food and Drug Administration, and the National Institute of Environmental Health Sciences have executed the Consortium Linking Academic and Regulatory Insights on BPA Toxicity (CLARITY-BPA) with Sprague Dawley rats.[42] This study was implemented with good laboratory practices to better illustrate the adverse health effects of BPA. The study outcomes suggest that BPA exposure could augment health complications by targeting multiple organ systems.[43] The following sections describe the association of BPA with pathological conditions as associated with metabolic, reproductive, developmental, neurological, cardiovascular, respiratory, and renal disorders, as well as cancer (Table 3.1).

Table 3.1 Molecular players in bisphenol A-associated human diseases.[a]

Diseases	Associated complications	Targets and signaling pathways	Effects	References
Reproductive diseases	Infertility		Reduces the number of fertilizing oocytes and lowers oocyte yield in women undergoing IVF	89
	Infertility	Fertility-related proteins	Reduces ATP level in spermatozoa Inhibits sperm motility and motion kinematics	90
	Reduced sperm parameters		Reduces sperm count, disrupts sperm morphology and motion, and induces DNA damage	91
	Reduced sperm quality	ERK	Negatively impacts sperm quality	92
	Endocrine disruption	ERK, genes involved in steroidogenesis	Decreases hormone production	93
	Ovary	Ki67 and caspase-3, PI3K/AKT pathway	Overactivates primordial follicles	94
	Ovarian insufficiency	PTEN	Premature activation of primordial follicles	95
	Miscarriage		Associated with immunoendocrine abnormalities and recurrent miscarriage	96
	Premature birth	PKC	Increases plasma concentrations of testosterone, estradiol and corticotrophin-releasing hormone	97
	PCOS	RXR/PPARγ pathway and FSHR Reduces GnRH	Alters steroidogenesis pathways Alters hypothalamic hormone action, estrus cyclicity	98, 99
Cardio-vascular diseases	Coronary artery disease	ERs, IKKs and CRP	Increases the risk of cardiovascular disease development	100
	Cardiac toxicity	TLR 2 and NF-κB	Induces DNA damage in cardiac cells	101
	Arrhythmias	PKA and CAMKII	Alters Ca^{2+} level and promotes arrhythmias in female cardiac myocytes	56
	Cardiac fibrosis	ERK 1/2	Induces cardiac fibroblasts proliferation and collagen production	102
	Blood pressure	ERα	Suppresses aortic ERα expression, increases systolic blood pressure in female offspring	103

Metabolic diseases	Obesity		Higher risk of obesity	104
	Obesity	Adipogenic signaling pathway	Induces adipogenic differentiation, dysregulation of metabolic homeostasis	105
	Insulin resistance	IL-6, IL-1B, TNFα, ADIPOQ and FABP4	Reduces insulin-stimulated glucose uptake	106
	Type I diabetes mellitus		Increases spontaneous insulitis and diabetes development in nonobese diabetic mice	107
	Type II diabetes mellitus	HbA1c	Elevated HbA1c in adults with type II diabetes mellitus	108
	Metabolism	PGC1α	Impairs cardiac mitochondrial function and interferes with the methylation of PGC1α	109
		MAPK pathway	Slows body metabolism and disrupts insulin signaling in mousee peripheral tissues	110
Chronic respiratory disease	Asthma		Increases odds of wheeze	72
		MAPK1	Alters MAPK1 methylation pattern	69
Developmental defects	Fertilization		Affects embryo cell number and embryo fragmentation score in IVF in males	111
	Ovary development	DNMT	Interferes with ovary development, induces reproductive injury, increases risk of abortion	112
	Placentation	Wnt/β-catenin pathway	Causes pregnancy failure and placental insufficiency	113
Thyroid disease	Neonatal thyroid function		Exposure during pregnancy shows reduced total T4 in women and decreased TSH in male neonates	114
	TH signaling		Dysregulates thyroid hormone level	115
	TH signaling		Interferes with TH signaling pathway that activates TH signaling in the absence of T3, but in the presence of TH shows antagonistic and/or agonistic actions	116
Chronic kidney disease	Nephrotoxicity		Differentially upregulates signaling lipids involved in inflammation and facilitates their localization in the renal cortex, medulla, and pelvis	117
	Kidney and renal function		Causes insufficient renal function	118
	Acute kidney injury		Impairs liver functional, induces oxidative stress, inflammation, and apoptosis	119

Table 3.1 (*Continued*)

Diseases	Associated complications	Targets and signaling pathways	Effects	References
Neurological disorders	Childhood behavior		Correlates with externalizing behaviors in 2-year-old children	120
	Developmental toxicity	KCC2	Disrupts KCC2 gene expression through epigenetic mechanisms	121
	Social behavior	SC1A1	Affects social interactions and behavior modification, especially in females	122
	Neurotoxicity	Wnt	Impairs hippocampus neurogenesis in developing rat brain	123
Hepatocellular diseases	NAFLD		Influences *de novo* fatty acid synthesis by increasing the expression of lipogenic genes	124
	Liver function		Induces oxidative stress, apoptosis, inflammatory response, and mitochondrial dysfunction in mice liver	125
Allergy and immune-related diseases	Allergic airway diseases		Induces parainflammation and airway allergic sensitization in mice	126
	Inflammation	p38 MAPK pathway	Induces allergic inflammation-related cytokines (TNF-α and IL-6) by blocking NF-κB and p38 kinase	127
	Inflammasome	IFN signaling	Induces innate immune sensors by stimulating type I IFN signaling and activates inflammasome activity	128

[a]AKT: serine/threonine protein kinase; ATP: adenosine triphosphate; CAMKII: Ca^{2+}/CaM-dependent protein kinase II; CRP: C-reactive protein; DNMT: DNA methyl transferase; ER: estrogen receptor; ERK: extracellular signal-regulated kinase; FSHR: follicle-stimulating hormone receptor; GnRH: gonadotropin-releasing hormone; IFN: interferon; IKK: IκB kinase; IL: interleukin; IVF: *in vitro* fertilization; MAPK: mitogen-activated protein kinase; NAFLD: non-alcoholic fatty liver disease; NF-κB: nuclear factor-κ light chain enhancer of activated B cells; PI3K: phosphatidylinositol 3 kinase; PKA: protein kinase A; PKC: protein kinase C; PPAR: peroxisome proliferator-activated receptor; PTEN: phosphatase and tensin homolog; RXR: retinoid X receptor; TH: thyroid hormone; TLR: Toll-like receptor; TNF: tumor necrosis factor; TSH: thyroid-stimulating hormone.

3.4.1 Obesity and Insulin Resistance

Studies have revealed the possible link between high serum or urine BPA levels and insulin resistance, obesity, and metabolic syndrome.[44] These metabolic disorders are indeed associated with signaling pathways related to inflammatory responses. BPA exposure increased serum levels of glucose, induced inflammation in pancreatic islet cells, and accelerated type-1 diabetes development. In rat INS-1 cells, BPA (0.020–2 μM) treatment impaired insulin secretion and induced apoptosis.[35] Elevated expression of pancreatic ERα/β, B-cell lymphoma 2 (BCL2), interleukin-1 (IL-1), and tumor necrosis factor-α (TNF-α) and high serum insulin levels were observed upon BPA exposure in the mouse model. A surge in glucose intolerance and serum TNF-α and IL-6 levels were reported with long-term oral exposure to BPA in mice.[45] Mice treated with 1 nM BPA displayed increased insulin release and glucose-stimulated Ca^{+2} signals, and reduced activity of ATP-sensitive potassium channel (KATP).[46] Exposure to BPA during pregnancy can impair insulin secretion and was correlated with the downregulation of cyclin-dependent kinase 4 (CDK4) and cyclin D2 (CCND2), reduced pancreatic β-cell mass, and enhanced expression of p53, cyclin-dependent kinase inhibitor 2A (CDKN2A) and β-cell apoptosis.[47] In matured adipocytes, the presence of BPA reduced AKT and ERK1/2 phosphorylation and insulin-stimulated glucose utilization.[48] BPA fosters adipogenesis and promotes genes involved in adipogenesis such as insulin-like growth factor 1 (IGF1), CCAAT-enhancer-binding proteins (C/EBPα), lipoprotein lipase (LPL), dual leucine zipper-bearing kinase (DLK), and PPARγ in human adipose tissue-derived stromal or stem cells.[49] Treatment of pre-adipocytes with 50 μM BPA upregulated genes involved in the mammalian target of rapamycin (mTOR) signaling pathway, adipogenesis, and lipid accumulation as analyzed by gene expression microarray.[50] Collectively, inflammatory responses induced by BPA are correlated with a higher risk of metabolic diseases including insulin resistance and obesity.

3.4.2 Cardiovascular Diseases

Exposure to BPA in the adult population is linked to a high risk of cardiovascular diseases such as heart attack, hypertension, and coronary and peripheral artery diseases.[51–53] Experimental evidence indicates that BPA exposure could interfere with cardiac functioning and foster aberrant cardiovascular activities, including atherosclerosis, arrhythmias, altered blood pressure, and cardiac remodeling.[53] Lang *et al.* showed a significant association between cardiovascular disease and high urinary BPA levels.[54] In female rodents, exposure to 10^{-9} M BPA during reperfusion may lead to a substantial increase in sustained ventricular arrhythmia duration. A low dose of BPA exposure to female rat cardiomyocytes triggered Ca^{2+}/CaM-dependent protein kinase II (CAMKII) and protein kinase A (PKA) signaling cascades. Elevated production of cAMP, phosphorylation of ryanodine receptor (RyR), stimulation of PKA, CAMKII, and phospholipase C (PLC), increased production of triphospho-inositol (IP_3), and Ca^{2+} release was observed upon acute BPA exposure.[55] Sex-specific variations in DNA methyltransferase levels, global DNA methylation, altered expression of Ca^{2+} handling proteins such as calsequestrin (CASQ2), sarco/endoplasmic reticulum Ca^{2+}-ATPase (SERCA), phospholamban (PLN) and sodium-calcium exchanger (NCX) was observed in the hearts of mice treated with BPA.[56] Low-dose BPA exposure was also shown to enhance atherosclerosis in mouse models.[57] The presence of BPA enhanced ROS production and mitochondrial dysfunction in the hearts of male rats as evidenced by diminished catalase activity, reduced GSH levels, and increased lipid peroxidation.[58] Altogether, the underlying mechanisms of BPA-mediated cardiac dysfunction may include variations in cardiac Ca^{2+} handling, ER signaling, genome or transcriptome alterations, and oxidative stress.

3.4.3 Neurological Disorders

Following BPA treatment, enhanced fear memory, migraine-like behaviors, dendritic morphogenesis of hippocampal neurons, and reduced spatial and visual memory have been reported.[59,60] Receptors such as ERα/β, *N*-methyl-D-aspartic acid (NMDA), insulin receptor (IR), γ-aminobutyric acid (GABA), GPR30 and glucocorticoid receptor (GR) are involved in the BPA-mediated neurological disorders.[12,61] Insulin receptor substrate proteins (IRS) and their downstream targets such as glycogen synthase kinase 3β (GSK3β) and AKT are stimulated by insulin binding to IR. These signal transducers are crucial for nervous system development. Glucose transporters (GLUTs) present in the blood–brain barrier mediate the transfer of glucose from the blood to the brain. Substantial upregulation of β-amyloid precursor protein (β-APP) and tau, reduced levels of IR, insulin, ERK1/2, AKT, GSK3β, and GLUT-1, -3, and -4 were observed in the brain upon perinatal BPA exposure.[62] BPA exposure suppressed the Wnt/β-catenin signaling pathway, inhibited neural stem cell proliferation and differentiation, and impaired myelination potential, survival, and differentiation of oligodendrocyte progenitor cells in the hippocampus, and thus led to reduced neurogenesis.[63] In the hippocampus, BPA treatment elevated autophagy protein expression, including beclin-1, microtubule-associated protein 1A/1B light chain 3B-II (MAP1LC3B-II), and high mobility group box 1 protein (HMGB1) with the subsequent downregulation of mTOR signaling and activation of AMP-activated protein kinase (AMPK), which in turn result in neurodegeneration in the hippocampus.[64] Continuous exposure to BPA (0.4, 4, or 40 mg kg^{-1} per day) for 90 days was shown to foster fear memory through enhanced histone acetylation and/or increased expression of NMDA receptors in the hippocampus.[65] In hippocampal neurons, BPA treatment augmented dendritic morphogenesis. These alterations are correlated with enhanced activation of ERK1/2 and increased levels of NMDA receptors and ER. The presence of BPA decreased RhoA expression, but enhanced RAC1/CDC42 expression in cultured hippocampal neurons.[66] El-Missiry *et al.*

reported cerebral cell apoptosis in rats exposed to 50 mg kg^{-1} of BPA, 3 days a week for 6 weeks. Elevated apoptosis was through caspase-3 and caspase-8 activation, enhanced oxidative stress and high levels of CD95 and p53 expression.[67] These data collectively suggest the potential of BPA in mediating neurological disorders.

3.4.4 Respiratory Disorders

BPA may increase the risk of wheezing, asthma, and bronchial hyper-responsiveness.[68,69] A study has shown increased levels of C–C motif chemokine ligand 2 (CCL2), Th2-relevant cytokines, IL-13, C–X–C motif chemokine ligand 1 (CXCL1), IL-6, and IL-1β in broncho-alveolar lavage fluid of male CD-1 mice treated with BPA (1 mg) and ovalbumin. High serum BPA was found to be associated with chronic obstructive pulmonary disease.[70] BPA was reported to decrease lung function with the increase in oxidative stress in aged people.[71] Prenatal exposure to BPA was shown to increase the odds of wheeze in young children.[72] In rhesus macaques, BPA treatment augmented secretory cell development in the proximal airways.[73] Vaidya and Kulkarni have revealed the correlation between allergic asthma and urinary BPA using National Health and Nutrition Examination Survey 2005–2006 data.[74] Additional investigations are essential to demonstrate the precise role of BPA in respiratory disorders.

3.4.5 Chronic Kidney Diseases

Association between low-grade albuminuria and high urine BPA has been demonstrated by epidemiological studies.[75] In cultured mouse podocytes, BPA was reported to induce hypertrophy and apoptosis.[76] Elevated proteinuria and hypertension were correlated with human exposure to BPA.[77] Treatment of rats with BPA (50, 100, and 150 mg kg^{-1}) for 5 weeks resulted in renal damage as evidenced by higher serum creatinine and urea and reduced creatinine clearance, glomerular injury, and proteinuria. Observed alterations were associated with reduced antioxidant superoxide dismutase and glutathione, increased lipid peroxidation, and mitochondrial dysfunction.[78]

3.4.6 Reproductive Disorders

The role of BPA in reproductive disorders has been extensively investigated.[12,79,80] BPA can control follicle development, oocyte maturation, and mammary gland formation in females. It can even affect fetal health through placental transfer.[80,81] In males, BPA can negatively affect spermatogenesis and sperm quality.[82] BPA (440 μM) exposure for 1 day was shown to stimulate the expression of DNA repair genes and proteins in the ovaries to address BPA-induced DNA damage.[83] Furthermore, BPA exposure induced premature apoptosis of oocytes. This is evident from studies that showed an elevation of the mRNA level of BCL2-associated X protein (BAX) in oocytes upon BPA treatment.[84] In female Sprague Dawley rats, BPA (0.001, 0.1 mg kg^{-1}) exposure for 90 days activated caspase-3 expression in the granulosa cells of the atretic follicles and the luteal cells of the corpus luteum. By the activation of caspase-3 and enhanced release of mitochondrial cytochrome, BPA was shown to enhance the injury of testicular germ cell line GC-2 cells derived from mouse spermatocytes.[85] A substantial increase in the levels of ATM serine/threonine kinase (ATM) and H2A histone family member X (γH2AX) was observed in male Wistar rats treated with 20 μg kg^{-1} BW per day of BPA. Elevated levels of DNA double-strand break markers are associated with meiotic abnormalities and spermiation delay, leading to impaired spermatogenesis.[86] BPA treatment of male Sprague Dawley rats reportedly enhanced Fas ligand (FasL), Fas, and caspase-3 expression in germ cells.[87] BPA reduces the nuclear translocation of androgen receptors by acting as an androgen receptor antagonist. BPA treatment can interfere with the hormone balance of testosterone, follicle-stimulating hormone (FSH), and luteinizing hormone (LH).[88]

3.5 BPA and Cancer

Cancer is another disease wherein BPA's role is extensively investigated. Despite these extensive investigations, the role of BPA as a cancer inducer is still controversial. BPA is implicated in the modulation of cancer cell survival, proliferation, growth, migration, invasion, therapeutic resistance, and apoptosis *via* distinct signaling cascades. BPA is reported to trigger several pro-carcinogenic signaling pathways by either binding to the membrane receptors or nuclear receptors.[40,129] Previous *in vivo* (Table 3.2) and *in vitro* (Table 3.3) studies have demonstrated the carcinogenic potential of BPA. Seachrist *et al.* have suggested categorizing BPA under Group 2A as a probable chemical carcinogen.[130] Growing evidence has revealed the ability of BPA to facilitate cancer hallmarks in different cell types, including breast, ovary, prostate, cervix, lung, colon, mesenchymal stem cells, and human trophoblasts.[43,130–135] BPA was shown to regulate key oncogenic signaling pathways such as epithelial–mesenchymal transition (EMT), Wnt, mitogen-activated protein kinase (MAPK), Janus kinase signal transducer and activator of transcription (JAK-STAT), NF-κB, cytokine signaling, senescence, apoptosis, Ca^{2+} homeostasis, and receptor-driven signaling cascades[130–134,136,137] (Figure 3.3). The following sections describe some of the molecular factors associated with BPA and carcinogenesis.

3.5.1 Epithelial–Mesenchymal Transition

EMT is a rescindable and dynamic cellular process opted by cancer cells to acquire more migratory and invasive properties during metastatic spread. EMT induction can foster cancer cell growth, survival, and therapeutic resistance. Many protein-coding genes and non-coding RNAs play a crucial role in EMT induction during carcinogenesis.[138] To acquire migratory,

Table 3.2 Molecular players in bisphenol A-mediated carcinogenesis—*in vivo* studies.[a]

Cell type	Study model	Dose and duration	Target genes and pathways	Effects	References
Ovary	BALB-C mice	100 $mg\,kg^{-1}$ BW, 15 weeks	↑ p-IRS and CCND1	↑ Tumor burden and expressions of ERα, pIRS-1, and CCND1 Crosstalk between ER and IGF-1R signals	148
Mammary gland	CD-1 mice	BPA *in utero* during gestation (days 9–21)	↓ IL-1β, IL-1RN, IL-7R, CXCL-2, CXCL-4, CCL-20, CXCL-14 and IRF-9	Dysregulation of inflammatory cytokines	149
	F433 rats	250 $mg\,kg^{-1}$ BW, PND 50	↑ IGF1R, RAF, ERK, ERB2, AKT, ERα and BCL2	↑ Proliferation and steroid receptor coactivators ↓ Apoptosis	150
Rat mammary gland	FVB/N-Tg mice	0, 50, 500 $ng\,kg^{-1}$ and 250 $\mu g\,kg^{-1}$ BW, gestational days 11–19	↑ ERα, p-ERα, CCND1, and c-Myc, ERBB2, EGFR, ERBB-3, ERK1/2 and AKT	↑ AKT-ERK signaling Promotes tumorigenesis	151
Resting mammary gland	Adult albino rats	5 $mg\,kg^{-1}$ daily for 8 weeks	↑ Ki-67 and caspase-3	Induces structural alterations Affected the proliferation rate ↑ Number and size of the acini and ducts in the mammary gland	152
Rat mammary gland	Sprague Dawley rats	250 $\mu g\,kg^{-1}$ BW, PND 50	Differential expression of AKT, caspases, PARP, ERα and ERβ	↑ Proliferation and risk of cancer	153
Rat mammary gland	C57BL/6 mice	250 $ng\,kg^{-1}$. embryonic day 19	Altered expression of genes involved in focal adhesion and adipogenesis pathways	↑ Maturation of the fat-pad and delays lumen formation in the epithelium	154
Human prostate epithelium	Nude mice	100 or 250 $\mu g\,kg^{-1}$ for 2 weeks	↑ AKT and ERK phosphorylation	↑ Hormone-dependent cancer risk ↑ Incidence of high-grade prostate intraepithelial neoplasia/cancer	155

[a]↑: upregulation; ↓: downregulation; AKT: serine/threonine protein kinase; BCL: B-cell lymphoma; BPA: bisphenol A; BW: body weight; CCND: cyclin D; EGFR: epidermal growth factor receptor; ER: estrogen receptor; ERK: extracellular signal-regulated kinase; IGF: insulin-like growth factor; IL: interleukin; ILR: interleukin receptor; IRF: interferon regulatory factor; IRS: insulin receptor substrate proteins; PARP: poly (ADP-ribose) polymerase; PND: post-natal day.

Table 3.3 Molecular players in BPA-mediated carcinogenesis—*in vitro* studies.[a]

Cancer type	Study model	Dose and duration	Target genes and pathways	Effects	References
Breast cancer	SkBr3 and MDA-MB-231 cells	10 nM for 48 hours	↓ mRNA and protein expression of FOXA1	↑ Migration and induces EMT	221
	MCF-7 cells	10 μM, 100 nM for 24–72 hours	↑ CXCR4, CCND1 and CCNE1 ↑ mRNA levels of ERRγ, PGC-1α, and PGC-1β	↑ Proliferation, migration, and invasion	144–147
	T-47D and MCF-7 cells	600 nM for 24 hours	↑ p53	↑ Proliferation and cellular viability	222
	MCF-7 cells	10 μM for 4 days	↑ miR-19a and miR-19b ↓ miR-19-PTEN, p-AKT, p-MDM2, p53, and PCNA	↑ Proliferation and cell cycle progression from G1 to S phase	223, 224
	T-47D and HRBEC cells	1 nM for 7 days	↓ p53, CDKN1A and BAX ↑ PCNA, cyclins, CDKs and p-pRb ↑ mTOR pathway proteins (AKT1, RPS6, pRPS6, and 4EBP1)	Blocks the pro-apoptotic effects of rapamycin ↑ Proliferation	187
	MCF-7 cells	0.1 mM for 24 hours	Activates TRPV1 channel, stimulates Ca^{2+} influx	↑ ROS, mitochondria oxidative stress, mitochondrial membrane depolarization	134
	MDA-MB-231	1 μM for 48 hours	Activates GPER-dependent pathway ↑ FAK, Src, and ERK2 and AP1/NF-κB DNA binding activity	↑ Migration and invasion	193
	MCF-7 cells	10, 100, or 1000 nM for 72 hours	↓ BRCA1 Activates ERα signaling pathways	↑ Proliferation	225

T-47D and MCF-7 cells	1, 2 μM for 24 hours	↑ p53	↑ Proliferation	222
SKBR3 cells	1 μM for 24 hours	↑ GPER target genes such as c-FOS, EGR-1, and CTGF through the PER/EGFR/ERK pathway	Stimulates cancer progression	174
MCF-7 cells	0.01–10 000 nM for 1, 3, and 24 hours	Activates PI3K/AKT and ERK signals *via* GPER	↑ Proliferation, intracellular ROS and Ca^{2+} influx	136
MCF-7 cells	4-methyl-2,4-bis-(4-hydroxy-phenyl) pent-1-ene (BPA metabolite)	↓ ERα protein expression Activates ERβ-mediated signaling	↑ Proliferation	226
MCF-7, MDA-MB-231 and SKBR3 cells	10 μM for 12 and 24 hours	↑ GPR30 and regulates the expression of c-fos likely through an AP1-mediated pathway	Activates ERK1/2 in both ERα/β-positive and negative cells	173
T47D and MDA-MB-468 cells	1 nM for 24 hours	↑ GPR30, ERRα, and ERR ↑ Anti-apoptotic BCL2 proteins	↓ Efficacy of chemotherapeutic agents	227
SkBr-3 and MDA-MB-231 cells	0.01–10 μM for 24 hours	Activates GPER/Cav-1/HSP90 signaling ↑ HIF-1α and VEGF	↑ Proliferative	228
MCF-7 cells	1 μM for 48 hours	↑ STAT3	↑ Proliferation	229
MCF-7 cells	10 μM for 24 hours	Regulate cell cycle and p38 MAPK pathway	↑ Proliferation	230
T47D and MCF7	10 nM to 10 μM for 24 hours and 7 days	↑ CXCL12 expression is mediated through ERα	↑ Proliferation	231
Human breast epithelial cell	1–10 mM for 2 weeks	↑ BRCA1, BRCA2, BARD1, CTIP, RAD51, and BRCC3	↑ Expressions of genes involved in DNA repair	232

Table 3.3 (*Continued*)

Cancer type	Study model	Dose and duration	Target genes and pathways	Effects	References
			↓ PDCD5 and BCL2L11 (BIM)		
	HMEC Cells	1, 10 μM for 1 week	↑ CDKN2A, p53, and CCNE	↑ Proliferation Hypermethylation of tumor suppressor genes (APC, BIN1, BMP6, BRCA1, CST6, ESRB, GSTP1, CDKN2A, CDKN1A, and TIMP3)	233
Ovarian cancer	BG-1 cells	1 μM for 0, 24, and 48 hours	↑ SNAI1, SLUG and VIM ↓ CDH1	↑ Migration	234
	SKOV3 cells	10, 100 nM for 24 hours	↑ MMP9 Activates canonical Wnt signaling	↑ EMT, migration, and invasion	235
	BG-1 cells		↑ VIM, CTSD, and MMP-2 ↓ CDH1 *via* ER signalling and TGF-β signaling	↑ EMT and migration	143
	OVCAR-3 cells	1 nM, 1 μM, 10 μM for 24 hours	↑ CD44 high CD24 low cell subpopulation ↑ ALDH1A1, CD133, SOX2, NANOG and OCT4	↑ Formation of stem cell sphere	162
	OVCAR-3 cells	1, 10, 40 and 100 nM for 72 hours	↑ p53 and ↓ caspase-3 and 7	↑ Proliferation	236
	SKOV-3 and OVCAR-3	1, 40 and 100 nM for 48 hours	↑ ERα, VEGF-R2 ↓ VEGF-A	Stimulates VEGF-R2 and VEGF-A expression	237
	BG-1 cells	1–10 μM	↑ SnoN, TGF-β pathway inhibitor, and c-Fos ↓ SMAD3	↑ Proliferation *via* inhibition of TGF-β signaling pathway	238
	BG-1 cells	10 μM for 0, 6 and 24 hours	↑ ERα and IGF-1R ↑ Phosphorylation of IRS-1 and AKT	↑ Proliferation by promoting the interaction between ERα and IGF-1R signaling pathways	148

	OVCAR-3 cells	1 nM, 1 pM for 24 hours	↑ ATP, lactate, and pyruvic acid	↑ Proliferation and glycolysis-based metabolism	133
	OVCAR-3 cells	0.2, 2, 8 and 20 ng mL^{-1}	↑ Phosphorylation of STAT3, ERK1/2, and AKT	↑ Leptin receptor expression and proliferation	170
	OVCAR-3 cells	40 nM for 24 hours	Inhibits caspase-3 expression *via* ERK1/2 pathway	Modulates STAT3 and ERK1/2 signaling pathways	239
	OVCAR-3 cells	0.1–100 nM	↑ MMP-2, MMP-9, and CDH2 Activates MAPK and PI3K/AKT signaling	↑ Migration	132
Prostate cancer	LNCaP cells	10, 50 and 100 µM	↓ CCND1 ↑ CDKN1A and 1B expressions ↑ p53	↓ Cell cycle progression	199
	PC-3 cells	1 and 10 µM for 96 hours	↑ Promoter DNA methylation of CDKN2A ↓ Expression of CDKN2A ↓ Chromatin-modifying enzymes KDM5B and NSD1	↓ CDKN2A ↑ Prostate cancer progression	200
	PC-3, LNCaP cells	0.1, 1, 10 and 100 nM for different time points	↑ CDK6 ↓ CDKN1A and 1B	Disrupt centrosome function and microtubule organization ↑ Anchorage-independent growth	198
	LNCaP cells	100 µM BPA for 24 hours	Activates ADAM17 ↑ Intracellular Ca^{2+}	↑ Apoptosis	240
	LNCaP cells	1–10 nM for 24 and 48 hours	↑ Store-Operated Calcium Entry (SOCE) ↑ORAI1 Activates potassium channel expression	↑ Migration	131
Colorectal cancer	SW480 cells	1 nM and 1 µM	↑ CDH2, SNAI1 ↓ CDH1	↑ Migration and invasion	241
	HCT116 cells	250 µM for 24 hours	↑ Caspase-3 and BAX ↑ Intracellular Ca^{2+} Activates MAPK/AKT signaling	↑ Mitochondrial dysfunction and apoptosis	181

Table 3.3 (*Continued*)

Cancer type	Study model	Dose and duration	Target genes and pathways	Effects	References
Lung cancer	A549 cells	10 μM BPA for 96 hours	↑ VIM/SNAI1/connexin (Cx)43 and ERRγ	↑ EMT and invasiveness	137
	A549 cells	10 μM for 0–72 hours	Activates GPER/EGFR/ERK1/2 signals ↑ MMPs	↑ *In vitro* migration and invasion	242
Endometrial cancer	RL95-2 cells	10 nM for 14 days	Activates MAPK pathway ↑ EMT and COX-2	↑ Migration and invasion	243
	HEC265 and Ishikawa cells	1 μM for 24 hours	↑ ERRγ target genes ↑ Ca^{2+} influx and EGF secretion Activates EGFR/ERK pathway	↑ Proliferation	175
Human mammary epithelial cells	HMEC cells	10 nM and 100 nM for 7 days	↑ CDKN2A and CCNE, BRCA1, CCNA1, CDKN2A, THBS1, TNFRSF10C, and TNFRSF10D	↑ Proliferation	233
Human bronchial epithelial cells	BEAS-2B cell	200 μM for 24 hours	↑ Caspase-3 Activates ATM/ATR/p53/γ-H2AX-dependent pathway	↑ Cytotoxicity, ROS, and DNA fragmentation	244
Papillary thyroid cancers	Nthy-ori3-1 cells	1 mM to 10 pM for 24, 48 and 72 hours	Activates ERK-COX2 signaling	↑ EMT, migration/invasion	245
Human thyroid papillary cancer	BHP10-3 cells	100 nM for 24 and 48 hours	↑ ERα/ERβ and GPR30 Phosphorylates AKT/mTOR	↑ Proliferation	172
Hemangioma	Hemangioma cells	10 nM for 48hours	↑ VIM ↓ CDH1	↑ Migration and invasion	246
Hepatocellular carcinoma	HepG2 cells	1 nM and 1 pM for 24, 48 and 72 hours	↑ Mitochondrial ROS Mitochondrial membrane potential hyperpolarization and pro-inflammatory cytokines secretion	↑ Mitochondrial dysfunction	216

Laryngeal squamous cell carcinoma	LSCC Hep-2, SNU899, and TU212 cells	10 nM for 48 hours	↑ MMP-2 and IL-6 Activates STAT3	↑ Proliferation, migration, and invasion	135
Cervical cancer	HeLa, SiHa, and C-33A cells	10 nM for 48 hours	↑ MMP-9 and fibronectin Activates NF-κB Phosphorylation of IKKβ *via* PKC signals	↑ Migration and invasion	194
Neuroblastoma	NB SH-SY5Y and SiMa cells	10 nM for 48 hours	↑ BCL2, VIM, and fibronectin and IL-6	↑ Proliferation, migration, and invasion	195
Acute myeloid leukemia	HL-60 and U937 cell	10 nM for 24 and 48 hours	Activates NF-κB ↓ IL-4 ↑ IL-6	↑ Proliferation	210

[a]↑: upregulation; ↓: downregulation AP1: activator protein 1; BAX: BCL2-associated X protein; BCL2: B-cell lymphoma 2; CCND1: cyclin D1; CCNE: cyclin E1; CDH1: E cadherin; CDH2: N cadherin; CDK: cyclin-dependent kinase; CTSD: cathepsin D; 4EBP1: eIF4E-binding protein 1; EGFR: epidermal growth factor receptor; EMT: epithelial–mesenchymal transition; ER: estrogen receptor; ERK: extracellular signal-regulated kinase; FAK: focal adhesion kinase; GPER: G protein-coupled estrogen receptor 1; GPR30: G-protein-coupled receptor 30; HRBEC: high-risk donor breast epithelial cells; IKK: IκB kinase; IL: interleukin; MAPK: mitogen-activated protein kinase; MMP: matrix metalloproteinase; NF-κB; nuclear factor-κ light chain enhancer of activated B cells; PCNA: proliferating cell nuclear antigen; PKC: protein kinase C; ROS: reactive oxygen species; RPS6: ribosomal protein S6; TGF: transforming growth factor; VEGF: Vascular endothelial growth factor; VIM: vimentin.

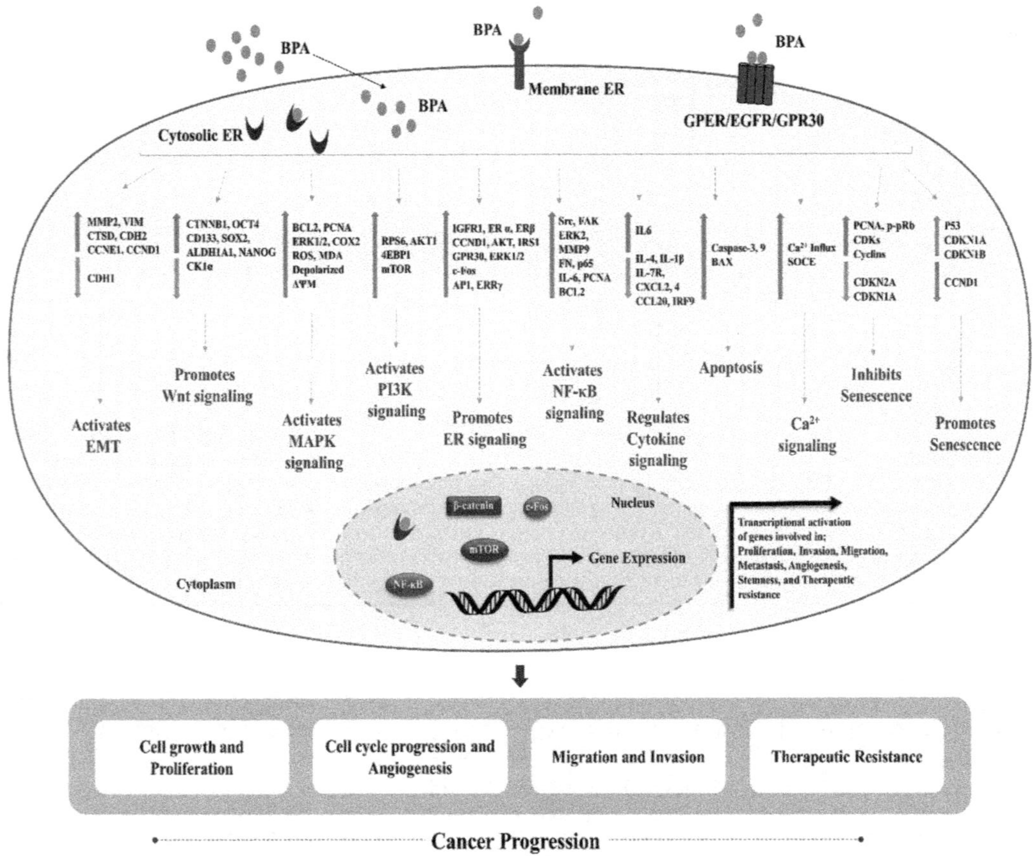

Figure 3.3 BPA-associated signaling pathways in cancer. BPA can promote neoplastic progression by altering the expression of different molecular players that are involved in key signaling pathways. Aberrant activation or repression of these signaling cascades results in enhanced proliferation, migration, invasion, and therapeutic resistance in cancer cells.

invasive, and mesenchymal phenotypes, epithelial cells exhibit a quasi-mesenchymal state during EMT.[139] Additionally, EMT critically regulates normal physiological and developmental processes such as tissue morphogenesis, gastrulation, wound healing, and neural crest formation.[140] Activation of several oncogenic signaling pathways including the ones driven by RAS, Src, Notch, integrins, Wnt/β-catenin, and growth factors such as hepatocyte growth factor (HGF), transforming growth factor (TGF), fibroblast growth factor (FGF), platelet-derived growth factor (PDGF), and epidermal growth factor receptor (EGFR) are well documented during EMT initiation.[141] Upon EMT induction, epithelial cells undergo molecular changes and morphological reorganization as evidenced by the reduced expression of epithelial markers such as desmoplakin (DSP), claudins (CLDNs), zona occludens-1 (ZO-1), E cadherin (CDH1), and occludin (OCLN), and elevated levels of mesenchymal markers including N cadherin (CDH2), vimentin (VIM), and EMT transcription factors (SNAI1/2 (snail family zinc finger 1/2), TWIST1/2 (twist family BHLH transcription factor 1/2), and ZEB1/2 (zinc finger E-box-binding homeobox 1/2)).[142] Being highly migratory and invasive, cells with a mesenchymal phenotype can evade apoptosis and exhibit resistance to chemotherapeutic drugs. The role of BPA in EMT induction has been demonstrated by several studies. For instance, A549 cells treated with BPA displayed morphological changes, higher invasiveness, E/N cadherin switch, upregulation of vimentin/snail-1/connexin(Cx)43 and enhanced drug resistance.[137] BPA exposure upregulated matrix metalloproteinase 2 (MMP2), VIM, cathepsin D (CTSD), reduced the expression of CDH1, and enhanced migratory potential through ER signaling in BG-1 ovarian cancer cells.[143] Higher proliferative and migratory ability, loss of cell–cell attachment, increased expression of cyclin E1 (CCNE), cyclin D1 (CCND1), and CDH2 and decreased levels of CDH1 were observed in ER-positive MCF7 clonal variant cells upon BPA treatment.[144–147] These data collectively indicate the EMT-inducing ability of BPA.

3.5.2 Wnt Signaling

The Wnt signaling cascade plays a critical role in several cellular events. Canonical, non-canonical and planar cell polarity Wnt pathways are stimulated by Wnt protein binding to Fizzled and LRP5/6 receptors. This binding activates disheveled, which in turn inhibits β-catenin (CTNNB1) degradation.[156] Consequently, CTNNB1 translocates to the nucleus from the cytosol and acts as a transcriptional activator for genes implicated in cell survival and proliferation. The absence of stimuli triggers proteasomal degradation of CTNNB1 through the collective action of the degradation complex in the cytosol.[157] Wnt signaling was initially reported for its ability to promote oncogenesis and then for its role in embryo development.[158,159] Furthermore, the Wnt cascade regulates tissue rejuvenation in the skin, intestine, and adult bone marrow.[160] BPA was shown to activate Wnt signaling in different cancer types. Hui *et al.* illustrated that low-dose BPA exposure significantly enhanced the nuclear translocation of CTNNB1 in ovarian cancer cells.[161] They concluded that BPA mediated alterations in gene expression and EMT induction through the canonical Wnt pathway. Another study has reported the activation of the WNT/β-catenin pathway, formation of a stem cell sphere, and augmented population of $CD44^{+}$, $CD24^{-}$ stem-like cells in OVCAR-3 cells.[162] These stem-like cells showed higher expression of octamer-binding transcription factor 4 (OCT4), CD133 (PROM1), SRY-Box Transcription Factor 2 (SOX2), aldehyde dehydrogenase 1 family member A1 (ALDH1A1), and NANOG, which are involved in tumor initiation and therapeutic resistance.[162] Treatment of nasopharyngeal carcinoma cells (NPC) with GSK-3β inhibitor restored BPA-induced cell proliferation. Mechanistically, BPA stabilized CTNNB1 mRNA by repressing the expression of miR-214-3p in NPC cells.[163] Furthermore, BPA-mediated reduction in CTNNB1 phosphorylation was due to reduced expression of casein kinase 1α (CK1α).[163] Thus, BPA can influence cancer cell proliferation, stemness, migration, and chemoresistance by regulating Wnt signaling.

3.5.3 Estrogen Receptor Signaling

ERs regulate cell differentiation, proliferation, migration, and apoptosis by acting as ligand-dependent transcription factors. ERα and ERβ are two distinct subtypes of ERs. The binding of estrogen to cytosolic receptors results in the dimerization of receptors followed by their translocation to the nucleus, wherein they interact with the estrogen response elements present in the target gene promoter.[164,165] Non-genomic estrogen functions are mediated by membrane-associated ERs, which include secondary messenger mobilization, interaction with other membrane receptors such as EGFR, and IGF1R and activation of effectors such as MAPK, phosphatidylinositol 3 kinase (PI3K), AKT, and Src.[166,167] Receptor-mediated responses to estrogen utilize membrane-associated ERs, whereas non-receptor-mediated actions are through ER-α36 and GPR30.[168,169] Impaired ER signaling upon BPA exposure is well documented in different cancers. BPA exposure promoted IGF1R and ERα expression, CCND1, AKT, and phosphorylation of IRS1 in BG-1-bearing mouse and ovarian cancer cells.[148,170] Upregulation of ERRγ is essential for BPA-induced breast cancer cell proliferation. ERK1/2 inhibitor treatment significantly inhibited breast cancer cell proliferation and reduced ERRγ levels.[171] These observations suggested the involvement of ERK1/2 in BPA-mediated ERRγ upregulation. In thyroid tumor cells, GPR30 and ERα/ERβ expression was enhanced upon BPA treatment.[172] Rapid activation of ERK1/2 after BPA treatment was observed in both ERα/β-negative and positive breast cancer cells.[173] GPR30 expression was critical for BPA-mediated transcriptional regulation of c-fos and ERK1/2 activation. Additionally, BPA modulated c-fos levels *via* an activator protein 1 (AP1)-mediated pathway in breast cancer cells.[173,174] Nuclear translocation of ERRγ and upregulation of BPA/ERRγ-target genes were reported in endometrial cancer cells treated with BPA.[175] These data indicate the role of BPA in derailing ER signaling.

3.5.4 MAPK Signaling

MAPK signaling has a central role in controlling cellular events such as differentiation, proliferation, and stress response. The binding of one or more growth factors to their respective receptors triggers transmembrane glycoproteins, which foster signal transduction *via* cytoplasmic intermediates leading to transcriptional or translational regulation of target genes.[176,177] MAPK, MAPK kinase, and MAPK kinase kinase are the three main kinases involved in the phosphorylation and activation of downstream targets.[178] p38MAPK, ERK1/2, ERK5, and JNK are the MAPK cascade members identified in eukaryotic cells. Studies have reported that the ERK/MAPK signaling pathway mainly regulates cell proliferation and differentiation, whereas p38MAPK and JNK pathways are involved in stress response and apoptosis.[176,179,180] Aberrant activation of MAPK signaling is associated with different pathological conditions, including cancer. BPA elevated intracellular and mitochondrial ROS, H_2O_2, and malondialdehyde levels, depolarized $\Delta\Psi M$ and induced apoptosis in colonic epithelial cells through MAPK and AKT pathways.[181] In MCF7 and SkBr3 breast cancer cells, nanomolar concentration of BPA significantly enhanced proliferative ability with the concomitant increase in the levels of BCL2, proliferating cell nuclear antigen (PCNA) and ERK1/2 phosphorylation.[171] BPA enhanced cell migration and COX2 expression, and activated MAPK by ERK1/2 phosphorylation in MDA-MB-231 and A549 cells.[182] Pre-treatment of OVCAR-3 cells with MAPK and PI3K inhibitors abolished the stimulatory effect of BPA on cell migration.[132] Taken together, BPA regulates key cancer hallmarks by modulating MAPK signaling.

3.5.5 PI3K/AKT/mTOR Signaling

PI3Ks belong to the family of lipid kinases and have the ability to phosphorylate inositol phospholipids to produce secondary messengers such as phosphatidylinositol-3,4,5-trisphosphate (PI-3,4,5-P(3)).[183] The receptor protein tyrosine kinase activation generates PIP3 and PIP2 at the inner membrane. Interaction of AKT with these phospholipids results in its translocation to the inner membrane, wherein it gets phosphorylated by phosphoinositide-dependent kinase 1 and 2.[184] AKT activation modulates different substrates, which take part in cell growth, survival, and cell cycle progression.[185,186] A growing literature suggests frequent alterations in the components of the PI3K pathway in human cancers. BPA exposure to high-risk donor breast epithelial cells (HRBECs) induced key genes and proteins of PI3K/AKT/mTOR pathway such as ribosomal protein S6 (RPS6), AKT1 and eukaryotic translation initiation factor 4E (eIF4E)-binding protein 1 (4EBP1) and tamoxifen resistance.[187,188] Elevated phosphorylation of AKT/mTOR was observed in human thyroid papillary cancer cells pretreated with BPA.[172]

3.5.6 NF-κB Signaling

NF-κB includes a group of five transcription factors that can bind to responsive gene promoters and regulate different cellular events. Several stimuli, including growth factors, ionizing radiation, cytokines, viral and bacterial products, ultraviolet radiation, DNA damage, oncogenic stress, and ROS can stimulate the activation of NF-κB.[189,190] These triggers can activate the IκB kinase (IKK) complex, which phosphorylates IκB, marking it for proteasomal degradation. Consequently, free NF-κB dimer relocates to the nucleus, where it regulates the transcription of target genes.[189,191,192] This well-defined signaling cascade has been shown to be dysregulated in cancer. In MDA-MB-cells, BPA exposure promoted cell migration by activating Src, focal adhesion kinase (FAK), ERK2, and AP1/NF-κB DNA binding ability.[193] Pre-treatment with BAY11-7082, a NF-κB inhibitor, substantially obliterated BPA-mediated upregulation of MMP9 and fibronectin in cervical cancer cell lines.[194] The rapid increase in the phosphorylation of p65 and IκBα and enhanced promoter activity of NF-κB was observed upon BPA treatment. Another study has shown that BPA treatment at nanomolar concentrations significantly enhanced the proliferative, migratory, and invasive ability of SiMa and SH-SY5Y cells.[195] These observations were further evidenced by elevated levels of IL-6, PCNA, and BCL2. BPA-mediated cell proliferation, invasion, and IL-6 expression were attenuated by NF-κB inhibitor treatment.[195] Furthermore, BPA exposure enhanced p65 phosphorylation, suggesting the ability of BPA to trigger neuroblastoma cell malignancy by activating NF-κB/IL-6 signals.[195]

3.5.7 Senescence

Senescence is a cell's intrinsic response to stress. Both tumor-suppressive and pro-oncogenic functions are linked to senescence. It functions as a tumor suppressor mechanism by inhibiting the replication of malignant cells.[196] In a senescent condition, cell cycle progression is inhibited by repression of proliferation enhancing genes such as E2F target gene and cyclin A and induction of senescence-associated heterochromatin foci.[197] Senescence induction upregulates CDKN2A and p53 expression in human primary cells. BPA-mediated upregulation of cell cycle inhibitors such as CDKN1A and CDKN1B, and the downregulation of CCND1 was observed in human prostate epithelial cells and prostate cancer cells.[198] In addition, EGFR-mediated phosphorylation of p53 at Ser15 was responsible for cell cycle arrest in BPA-treated primary and prostate cancer cells.[199] In PC3 cells, a substantial increase in promoter DNA methylation and reduced expression of CDKN2A was observed after 96 hours of BPA (1 μM, 10 μM) exposure. Chromatin immunoprecipitation results revealed a significant increase in H3K9me3, H3K9ac, and H3K27me3 modifications in CDKN2A gene.[200] BPA exposure in T47D and HRBEC cell lines exhibited a marked reduction in the expression of cell cycle regulators such as CDKN1A, p53, and BAX with the subsequent increase in pro-proliferative genes including PCNA, p-pRb, CDKs, and cyclins.[187] Collectively, these studies suggest the senescence inducing as well as inhibiting properties of BPA in cancer models.

3.5.8 Apoptosis

A mechanism of programmed cell death called apoptosis has an important role in the growth and developmental processes. Two different signaling pathways that are involved in cellular apoptosis have been described: extrinsic and intrinsic apoptotic pathways.[201] Extracellular signals such as death-inducing signals from cytotoxic T cells and intracellular signals such as cytokine and growth factor deprivation and DNA damage can trigger apoptotic signaling.[202] Apoptotic stimuli activate caspases that cleave target proteins. Caspase-2, -8, -9, and 10 are called initiator caspases, whereas caspase-3, -6, and -7 serve as executioner caspases, which eventually cleave the target proteins and induce cell death.[203] The apoptotic cascades are tightly controlled by the combined activities of anti-apoptotic proteins, including BCL2 and proapoptotic proteins such as BAX and BCL2 homologous antagonist killer (BAK).[204] Deregulated apoptosis is an indicator of various disorders. Elevated apoptosis has been observed in acute and chronic degenerative and immunodeficiency disorders and infertility.[205] Apoptotic delays or inhibition have been demonstrated in autoimmunity and cancer. BPA treatment reduced the viability of human promyelocytic leukemia (HL-60), human submandibular gland, and human oral squamous cell carcinoma cells (HSC-2). Caspase-3 and -9 activation was seen after BPA treatment in HL-60 cells.[206] BPA stimulated inter-nucleosomal DNA fragmentation, which in turn is a biomarker for apoptosis.[206] Presence of BPA reduced the proliferation and viability of colonic epithelial cells through upregulation of BAX and caspase-3 and apoptosis induction.[181]

3.5.9 Cytokines

Cytokines produced by the adaptive and innate immune systems cells regulate immune system homeostasis by acting as mediators of intracellular signaling.[207] These are membrane-bound or secreted proteins produced in response to infection or inflammation or tumor antigens.[208] Following cytokine binding, high-affinity receptors on the target cell membrane trigger the intracellular signaling cascade, modulating gene transcription. Failure in resolving tissue injury can result in continuous cytokine production and aggravation of tissue destruction.[209] In BPA-treated acute myeloid leukemia (AML) cells, reduced levels of IL-4 and higher levels of IL-6 were reported. Neutralization antibody of IL-6 or ectopic expression of IL-4 attenuated BPA-mediated cell proliferation and chemoresistance in AML cells.[210] Furthermore, BPA-induced IL-6 upregulation was through NF-κB activation in AML cells.[210] Substantial reduction in the expression of IL-1β, IL-1RN, IL-7 receptor, CXCL-2, CXCL-4, chemokine ligand-20 (CCL-20), CXCL-14, interferon regulatory factor (IRF-9), and immune responsive gene (IRG-1) was observed after BPA exposure *in utero*.[149] This study suggested that dysregulation of both anti-inflammatory and pro-inflammatory cytokines results in a microenvironment that can foster cancer progression.[149]

3.5.10 BPA and Mitochondria

Mitochondria play a central role in maintaining normal cellular metabolism.[211] They are involved in several cellular functions, including Ca^{2+} homeostasis, fatty acid and amino acid metabolism, cell growth, ROS generation, and cell death.[212] Metabolic alterations in mitochondria can lead to severe health complications such as diabetes, muscular dystrophy, autism, cardiovascular disorders, and multiple cancers.[213–215] Growing bodies of evidence suggest the potential of BPA in inducing molecular, structural, and functional alterations in mitochondria. BPA modulated transient receptor potential vanilloid 1 (TRPV1) channel activity, enhanced cell proliferation and mitochondrial oxidative stress in MCF7 breast cancer cells.[134] In HepG2 cells, BPA exposure substantially induced mitochondrial dysfunction as evidenced by lipid accumulation, $\Delta\Psi M$ hyperpolarization, lipoperoxidation, ROS production, and release of pro-inflammatory cytokines.[216] Intraperitoneal administration of 0.05 and 1.2 $mg\,kg^{-1}$ BW per day of BPA impaired hepatic mitochondria structure, with enhanced MDA, TNF-α, and IL-6 levels and reduced levels of glutathione peroxidase-3 in mice.[217] BPA exposure declined the rate of oxygen consumption, $\Delta\Psi M$, and ATP production in HepG2 cells.[217] In neuroblastoma cells, BPA was shown to diminish superoxide dismutase and MDA levels, $\Delta\Psi M$, and BCL2 expression with the consequent upregulation of BAX, BAK1, caspase-3, and cytochrome c proteins. Colonic epithelial cells displayed depolarized $\Delta\Psi M$ and loss of mitochondrial integrity upon BPA exposure.[218]

3.5.11 Calcium Signaling

A ubiquitous intracellular secondary messenger, Ca^{2+}, critically regulates several physiological functions and biological processes. Specific stimuli can elevate intracellular Ca^{2+} levels through Ca^{2+} entry from the extracellular space and release of Ca^{2+} from the endoplasmic reticulum. Ca^{2+} pumps, channels, and exchangers modulate cytosolic and organellar Ca^{2+} levels.[219,220] Owing to its distinct role, derailed Ca^{2+} homeostasis has been well demonstrated in different pathological conditions. Towards this, few studies have investigated the effect of BPA on Ca^{2+} signaling. BPA treatment fostered Ca^{2+} influx in Ishikawa cells.[175] BPA-mediated modulation of TRPV1 channel activity was shown in MCF7 breast cancer cells.[134] Pretreatment of LNCaP cells with BPA amplified store-operated calcium entry (SOCE).[131] BPA-induced SOCE amplification was found to be dependent on ORAI1 (calcium release-activated calcium channel protein 1).[131] Furthermore, BPA-mediated prostate cancer cell migration was reduced upon extracellular Ca^{2+} chelation or upon treatment with SOCE inhibitors such as BTP2 and SKF96365.[131]

3.6 Conclusions and Future Perspectives

BPA exerts its toxicological effects by modulating the intricate cellular signaling network. By inducing genetic and epigenetic modifications, endocrine disruption, oxidative stress, mitochondrial dysfunction, and interacting with key signaling molecules, BPA can induce adverse health effects in humans. A plethora of evidence from epidemiological studies, *in vitro* and *in vivo* studies have convincingly demonstrated the potency of BPA in causing different chronic health complications such as obesity, diabetes, reproductive, developmental, cardiovascular, and neurological disorders, as well as cancer. Data from various studies have illustrated the molecular mechanism of BPA action during neoplastic progression. Key signaling pathways such as EMT, Wnt, PI3K/AKT, MAPK, ER, NF-κB, cytokines, senescence, apoptosis, Ca^{2+} homeostasis, and mitochondrial functionalities are substantially altered through BPA. In this regard, more systematic and comprehensive functional investigations are necessitated to uncover the molecular complexity involved in BPA mediated toxicity. Instant efforts are needed to explore environmentally safe, non-toxic and economical substitutes for BPA. Since this is an ever-evolving area of research, investigations on BPA-driven organellar toxicity, epigenetic modifications, and better approaches to mitigate the toxic effects of BPA could be attempted to either prevent or reduce its adverse health effects.

Acknowledgements

We thank ICMR-Senior Research Fellowship (Reference ID-2019/4115/CMB/BMS and 2020/8704/CMB/BMS), Government of India, Directorate of Minorities Fellowship (DOM/FELLOWSHIP/CR-10/2019-20), DST-Ph.D. fellowship, KSTePS, Department of Science and Technology (DST), Government of Karnataka (Reference ID- DST/KSTePS/Ph.D. Fellowship/LIF-11: 2019-20) and Dr TMA Pai Structured Ph.D. Fellowship, Manipal Academy of Higher Education (MAHE) for financial assistance. We thank Manipal Academy of Higher Education, Manipal, for infrastructure support. We acknowledge the Center for DNA Repair and Genome Stability (CDRGS) at the Manipal Academy of Higher Education.

References

1. C. Campanale, C. Massarelli, I. Savino, V. Locaputo and V. F. Uricchio, A detailed review study on potential effects of microplastics and additives of concern on human health, *Int. J. Environ. Res. Public Health*, 2020, **17**, 1212.

2. J. Corrales, L. A. Kristofco and W. B. Steele, *et al.*, Global Assessment of Bisphenol A in the Environment, *Dose–Response*, 2015, 1–29.
3. A. L. Blankenship and K. Coady, Bisphenol A, in *Encyclopedia of Toxicology*, 2005.
4. P. Allard and M. P. Colaiácovo, Bisphenol A, in *Reproductive and Developmental Toxicology*, 2011.
5. D. Chen, K. Kannan and H. Tan, *et al.*, Bisphenol Analogues Other Than BPA: Environmental Occurrence, Human Exposure, and Toxicity - A Review, *Environ. Sci. Technol.*, 2016, 5438–5453.
6. S. Almeida, A. Raposo, M. Almeida-González and C. Carrascosa, Bisphenol A: Food Exposure and Impact on Human Health, *Compr. Rev. Food Sci. Food Saf.*, 2018, 1503–1517.
7. F. Acconcia, V. Pallottini and M. Marino, Molecular mechanisms of action of BPA, *Dose–Response*, 2015, 1–9.
8. L. N. Vandenberg, M. V. Maffini, C. Sonnenschein, B. S. Rubin and A. M. Soto, Bisphenol-A and the Great Divide: A Review of Controversies in the Field of Endocrine Disruption, *Endocr. Rev.*, 2009, **30**(1), 75–95.
9. L. N. Vandenberg, T. Colborn and T. B. Hayes, *et al.*, Hormones and endocrine-disrupting chemicals: Low-dose effects and nonmonotonic dose responses, *Endocrine Reviews*, 2012, 378–455.
10. L. N. Vandenberg, Non-monotonic dose responses in studies of endocrine disrupting chemicals: Bisphenol a as a case study, *Dose–Response*, 2014, 259–276.
11. F. Liguori, C. Moreno-Marrodan and P. Barbaro, Biomass-derived chemical substitutes for bisphenol A: Recent advancements in catalytic synthesis, *Chem. Soc. Rev.*, 2020, 6329–6363.
12. R. Rezg, S. El-Fazaa, N. Gharbi and B. Mornagui, Bisphenol A and human chronic diseases: Current evidences, possible mechanisms, and future perspectives, *Environ. Int.*, 2014, 83–90.
13. J. E. Cooper, E. L. Kendig and S. M. Belcher, Assessment of bisphenol A released from reusable plastic, aluminium and stainless steel water bottles, *Chemosphere*, 2011, 943–947.
14. M. R. Bernier and L. N. Vandenberg, Handling of thermal paper: Implications for dermal exposure to bisphenol A and its alternatives, *PLoS One*, 2017, e0178449.
15. EFSA Panel on food contact materials, enzymes, flavourings and processing aids (CEF), Scientific Opinion on Bisphenol A: evaluation of a study investigating its neurodevelopmental toxicity, review of recent scientific literature on its toxicity and advice on the Danish risk assessment of Bisphenol A, *EFSA J.*, 2010, **8**(9), 1829.
16. C. W. Choi, J. Y. Jeong and M. S. Hwang, *et al.*, Establishment of the korean tolerable daily intake of bisphenol a based on risk assessments by an expert committee, *Toxicol. Res.*, 2013, 285–291.
17. S. J. Genuis, S. Beesoon, D. Birkholz and R. A. Lobo, Human Excretion of Bisphenol A: Blood, Urine, and Sweat (BUS) Study, *J. Environ. Public Health*, 2012, **2012**, 1–10.

18. R. Valentino, V. D'Esposito and F. Ariemma, *et al.*, Bisphenol A environmental exposure and the detrimental effects on human metabolic health: is it necessary to revise the risk assessment in vulnerable population?, *J. Endocrinol. Invest.*, 2016, **39**(3), 259–263.
19. G. K. Prasanth, L. M. Divya and C. Sadasivan, Bisphenol-A can bind to human glucocorticoid receptor as an agonist: An in silico study, *J. Appl. Toxicol.*, 2010, 769–774.
20. M. Murata and J. H. Kang, Bisphenol A (BPA) and cell signaling pathways, *Biotechnol. Adv.*, 2018, 311–327.
21. R. Meli, A. Monnolo, C. Annunziata, C. Pirozzi and M. C. Ferrante, Oxidative Stress and BPA Toxicity: An Antioxidant Approach for Male and Female Reproductive Dysfunction, *Antioxidants*, 2020, **9**(5), 405.
22. J. Peretz, L. Vrooman and W. A. Ricke, *et al.*, Bisphenol A and reproductive health: Update of experimental and human evidence, 2007-2013, *Environ. Health Perspect.*, 2014, 775–786.
23. P. Allard and M. P. Colaiácovo, Bisphenol A impairs the double-strand break repair machinery in the germline and causes chromosome abnormalities, *Proc. Natl. Acad. Sci. U. S. A.*, 2010, 20405–20410.
24. G. S. Prins, W. Y. Tang, J. Belmonte and S. M. Ho. Perinatal exposure to oestradiol and bisphenol A alters the prostate epigenome and increases susceptibility to carcinogenesis. in *Basic and Clinical Pharmacology and Toxicology*, 2008, 134–138.
25. J. G. Bromer, Y. Zhou, M. B. Taylor, L. Doherty and H. S. Taylor, Bisphenol-A exposure in utero leads to epigenetic alterations in the developmental programming of uterine estrogen response, *FASEB J.*, 2010, 2273–2280.
26. Y. I. Weng, P. Y. Hsu and S. Liyanarachchi, *et al.*, Epigenetic influences of low-dose bisphenol A in primary human breast epithelial cells, *Toxicol. Appl. Pharmacol.*, 2010, 111–121.
27. K. Warita, T. Mitsuhashi and K. ichi Ohta, *et al.*, Gene expression of epigenetic regulatory factors related to primary silencing mechanism is less susceptible to lower doses of bisphenol a in embryonic hypothalamic cells, *J. Toxicol. Sci.*, 2013, 285–289.
28. L. F. Doherty, J. G. Bromer, Y. Zhou, T. S. Aldad and H. S. Taylor, In utero exposure to diethylstilbestrol (DES) or bisphenol-A (BPA) increases EZH2 expression in the mammary gland: An epigenetic mechanism linking endocrine disruptors to breast cancer, *Horm. Cancer*, 2010, 146–155.
29. E. Dhimolea, P. R. Wadia and T. J. Murray, *et al.*, Prenatal exposure to BPA alters the epigenome of the rat mammary gland and increases the propensity to neoplastic development, *PLoS One*, 2014, 1–9.
30. B. De Felice, F. Manfellotto and A. Palumbo, *et al.*, Genome-wide microRNA expression profiling in placentas from pregnant women exposed to BPA, *BMC Med. Genomics*, 2015, 1–3.
31. A. Veiga-Lopez, L. J. Luense, L. K. Christenson and V. Padmanabhan, Developmental programming: Gestational bisphenol-A treatment alters trajectory of fetal ovarian gene expression, *Endocrinology*, 2013, 1873–1884.

32. M. Schieber and N. S. Chandel, ROS function in redox signaling and oxidative stress, *Curr. Biol.*, 2014, 453–462.
33. Y. J. Yang, Y. C. Hong and S. Y. Oh, *et al.*, Bisphenol A exposure is associated with oxidative stress and inflammation in postmenopausal women, *Environ. Res.*, 2009, 797–801.
34. H. Gong, X. Zhang and B. Cheng, *et al.*, Bisphenol A Accelerates Toxic Amyloid Formation of Human Islet Amyloid Polypeptide: A Possible Link between Bisphenol A Exposure and Type 2 Diabetes, *PLoS One*, 2013, e460.
35. Y. Lin, X. Sun and L. Qiu, *et al.*, Exposure to bisphenol A induces dysfunction of insulin secretion and apoptosis through the damage of mitochondria in rat insulinoma (INS-1) cells, *Cell Death Dis.*, 2013, **4**(1), e460.
36. E. Diamanti-Kandarakis, J. P. Bourguignon and L. C. Giudice, *et al.*, Endocrine-disrupting chemicals: An Endocrine Society scientific statement, *Endocr. Rev.*, 2009, 293–342.
37. I. Cimmino, F. Fiory, G. Perruolo, C. Miele, F. Beguinot, P. Formisano and F. Oriente, Potential Mechanisms of Bisphenol A (BPA) Contributing to Human Disease, *Int. J. Mol. Sci.*, 2020, **21**, 5761.
38. H. Masuyama and Y. Hiramatsu, Involvement of Suppressor for Gal 1 in the Ubiquitin/Proteasome-mediated Degradation of Estrogen Receptors, *J. Biol. Chem.*, 2004, 12020–12026.
39. S. Lee, K. Suk and I. K. Kim, *et al.*, Signaling pathways of bisphenol A-induced apoptosis in hippocampal neuronal cells: Role of calcium-induced reactive oxygen species, mitogen-activated protein kinases, and nuclear factor-κB, *J. Neurosci. Res.*, 2008, 2932–2942.
40. S. Nomiri, R. Hoshyar, C. Ambrosino, C. R. Tyler and B. Mansouri, A mini review of bisphenol A (BPA) effects on cancer-related cellular signaling pathways, *Environ. Sci. Pollut. Res.*, 2019, **26**(9), 8459–8467.
41. R. Viñas and C. S. Watson, Mixtures of xenoestrogens disrupt estradiol-induced non-genomic signaling and downstream functions in pituitary cells, *Environ. Health: Global Access Sci. Source*, 2013, 26.
42. J. J. Heindel, R. R. Newbold and J. R. Bucher, *et al.*, NIEHS/FDA CLARITY-BPA research program update, *Reprod. Toxicol.*, 2015, **58**, 33–44.
43. G. S. Prins, H. B. Patisaul, S. M. Belcher and L. N. Vandenberg, CLARITY-BPA academic laboratory studies identify consistent low-dose Bisphenol A effects on multiple organ systems, *Basic Clin. Pharmacol. Toxicol.*, 2019, **125**(S3), 14–31.
44. L. Caporossi and B. Papaleo, Bisphenol A and Metabolic Diseases: Challenges for Occupational Medicine, *Int. J. Environ. Res. Public Health*, 2017, **14**, 959.
45. M. K. Moon, I. K. Jeong and T. J. Oh, *et al.*, Long-term oral exposure to bisphenol A induces glucose intolerance and insulin resistance, *J. Endocrinol.*, 2015, 35–42.
46. S. Soriano, A. B. Ropero and P. Alonso-Magdalena, *et al.*, Rapid regulation of KATP channel activity by 17β-estradiol in pancreatic β-cells involves the estrogen receptor β and the atrial natriuretic peptide receptor, *Mol. Endocrinol.*, 2009, 1973–1982.

47. P. Alonso-Magdalena, M. García-Arévalo, I. Quesada and Á. Nadal, Bisphenol-A Treatment During Pregnancy in Mice: A New Window of Susceptibility for the Development of Diabetes in Mothers Later in Life, *Endocrinology*, 2015, **156**(5), 1659–1670.
48. F. Ariemma, V. D'Esposito and D. Liguoro, *et al.*, Low-Dose bisphenol-A impairs adipogenesis and generates dysfunctional 3T3-L1 adipocytes, *PLoS One*, 2016, e0150762.
49. K. Zhang, F. Wang and J. Huang, *et al.*, Insulin-like growth factor 2 promotes the adipogenesis of hemangioma-derived stem cells, *Exp. Ther. Med.*, 2018, 1663–1669.
50. J. G. Boucher, M. Husain and A. Rowan-Carroll, *et al.*, Identification of mechanisms of action of bisphenol A-induced human preadipocyte differentiation by transcriptional profiling, *Obesity*, 2014, 2333–2343.
51. S. G. Chrysant, Association of Exposure to Bisphenol A and Incidence of Cardiovascular Disease and Hypertension, *J. Clin. Hypertens.*, 2015, 737–739.
52. C. Han and Y. C. Hong, Bisphenol A, Hypertension, and Cardiovascular Diseases: Epidemiological, Laboratory, and Clinical Trial Evidence, *Curr. Hypertens. Rep.*, 2016, 11.
53. X. Gao and H. S. Wang, Impact of bisphenol A on the cardiovascular system - Epidemiological and experimental evidence and molecular mechanisms, *Int. J. Environ. Res. Public Health*, 2014, 8399–8413.
54. I. A. Lang, T. S. Galloway and A. Scarlett, *et al.*, Association of Urinary Bisphenol A concentration with medical disorders and laboratory abnormalities in adults, *JAMA, J. Am. Med. Assoc.*, 2008, 1303–1310.
55. X. Gao, J. Ma, Y. Chen and H. S. Wang, Rapid responses and mechanism of action for low-dose bisphenol S on ex vivo rat hearts and isolated myocytes: Evidence of female-specific proarrhythmic effects, *Environ. Health Perspect.*, 2015, 571–578.
56. X. Gao, Q. Liang, Y. Chen and H. S. Wang, Molecular mechanisms underlying the rapid arrhythmogenic action of bisphenol A in female rat hearts, *Endocrinology*, 2013, 4607–4617.
57. Y. Sui, S. Park and R. N. Helsley, *et al.*, Bisphenol A Increases Atherosclerosis in Pregnane X Receptor-Humanized ApoE Deficient Mice, *J. Am. Heart Assoc.*, 2014, **3**(2), e000492.
58. H. S. Aboul Ezz, Y. A. Khadrawy and I. M. Mourad, The effect of bisphenol A on some oxidative stress parameters and acetylcholinesterase activity in the heart of male albino rats, *Cytotechnology*, 2015, 145–155.
59. F. Xin, E. Fischer and C. Krapp, *et al.*, Mice exposed to bisphenol A exhibit depressive-like behavior with neurotransmitter and neuroactive steroid dysfunction, *Horm. Behav.*, 2018, 93–104.
60. E. Kimura, C. Matsuyoshi and W. Miyazaki, *et al.*, Prenatal exposure to bisphenol A impacts neuronal morphology in the hippocampal CA1 region in developing and aged mice, *Arch. Toxicol.*, 2016, 691–700.
61. I. Cimmino, F. Fiory and G. Perruolo, *et al.*, Potential mechanisms of bisphenol a (BPA) contributing to human disease, *Int. J. Mol. Sci.*, 2020, **21**(16), 1–22.

62. J. Li, Y. Wang and F. Fang, *et al.*, Bisphenol A disrupts glucose transport and neurophysiological role of IR/IRS/AKT/GSK3β axis in the brain of male mice, *Environ. Toxicol. Pharmacol.*, 2016, 7–12.
63. S. K. Tiwari, S. Agarwal and B. Seth, *et al.*, Inhibitory Effects of Bisphenol-A on Neural Stem Cells Proliferation and Differentiation in the Rat Brain Are Dependent on Wnt/β-Catenin Pathway, *Mol. Neurobiol.*, 2015, 1735–1757.
64. S. Agarwal, S. K. Tiwari and B. Seth, *et al.*, Activation of Autophagic Flux against Xenoestrogen Bisphenol-A-induced Hippocampal Neurodegeneration via AMP kinase (AMPK)/Mammalian Target of Rapamycin (mTOR) Pathways, *J. Biol. Chem.*, 2015, **290**(34), 21163–21184.
65. Q. Zhang, X. Xu and T. Li, *et al.*, Exposure to bisphenol-A affects fear memory and histone acetylation of the hippocampus in adult mice, *Horm. Behav.*, 2014, 106–113.
66. X. Xu, Y. Lu and G. Zhang, *et al.*, Bisphenol A promotes dendritic morphogenesis of hippocampal neurons through estrogen receptor-mediated ERK1/2 signal pathway, *Chemosphere*, 2014, 129–137.
67. M. A. El-Missiry, A. I. Othman, M. A. Al-Abdan and A. A. El-Sayed, Melatonin ameliorates oxidative stress, modulates death receptor pathway proteins, and protects the rat cerebrum against bisphenol-A-induced apoptosis, *J. Neurol. Sci.*, 2014, **347**(1–2), 251–256.
68. L. Robinson and R. Miller, The Impact of Bisphenol A and Phthalates on Allergy, Asthma, and Immune Function: a Review of Latest Findings, *Curr. Environ. Health Rep.*, 2015, 379–387.
69. C.-F. Yang, W. J. J. Karmaus, C.-C. Yang, M.-L. Chen and I.-J. Wang, Bisphenol a Exposure, DNA Methylation, and Asthma in Children, *Int. J. Environ. Res. Public Health*, 2020, **17**, 298.
70. M. He, T. Ichinose and S. Yoshida, *et al.*, Exposure to bisphenol A enhanced lung eosinophilia in adult male mice, *Allergy, Asthma, Clin. Immunol.*, 2016, 1–9.
71. J. H. Kim, S. Bae, K. Lee, J.-R. Sohn and Y.-C. Hong, Exposure to Bisphenol A and Phthalates Affects Lung Function and Oxidative Stress in the Elderly, *Epidemiol.*, 2009, S154.
72. A. J. Spanier, R. S. Kahn and A. R. Kunselman, *et al.*, Prenatal exposure to bisphenol A and child wheeze from birth to 3 years of age, *Environ. Health Perspect.*, 2012, 916–920.
73. L. S. Van Winkle, S. R. Murphy, M. V. Boetticher and C. A. VandeVoort, Fetal exposure of rhesus macaques to bisphenol a alters cellular development of the conducting airway by changing epithelial secretory product expression, *Environ. Health Perspect.*, 2013, 912–918.
74. S. V. Vaidya and H. Kulkarni, Association of urinary bisphenol a concentration with allergic asthma: Results from the national health and nutrition examination survey 2005–2006, *J. Asthma*, 2012, 800–806.
75. L. Trasande, T. M. Attina and H. Trachtman, Bisphenol A exposure is associated with low-grade urinary albumin excretion in children of the United States, *Kidney Int.*, 2013, 741–748.

76. N. Olea-Herrero, M. I. Arenas and C. Muñóz-Moreno, *et al.*, Bisphenol-A induces podocytopathy with proteinuria in mice, *J. Cell. Physiol.*, 2014, 2057–2066.
77. R. J. Bosch, B. Quiroga and C. Muñoz-Moreno, *et al.*, Bisphenol A: An environmental factor implicated in renal vascular damage, *Nefrologica (Engl. Edn.)*, 2016, 5–9.
78. A. Kobroob, W. Peerapanyasut, N. Chattipakorn and O. Wongmekiat, Damaging Effects of Bisphenol A on the Kidney and the Protection by Melatonin: Emerging Evidences from In Vivo and In Vitro Studies, *Oxid. Med. Cell. Longevity*, 2018, **2018**, 1–15.
79. E. Matuszczak, M. D. Komarowska, W. Debek and A. Hermanowicz, The Impact of Bisphenol A on Fertility, Reproductive System, and Development: A Review of the Literature, *Int. J. Endocrinol.*, 2019, **2019**, 1–8.
80. X. Huo, D. Chen and Y. He, *et al.*, Bisphenol-a and female infertility: A possible role of gene-environment interactions, *Int. J. Environ. Res. Public Health*, 2015, 11101–11116.
81. M. Nishikawa, H. Iwano and R. Yanagisawa, *et al.*, Placental transfer of conjugated bisphenol A and subsequent reactivation in the rat fetus, *Environ. Health Perspect.*, 2010, 1196–1203.
82. C. B. Herath, W. Jin and G. Watanabe, *et al.*, Adverse effects of environmental toxicants, octylphenol and bisphenol A, on male reproductive functions in pubertal rats, *Endocrine*, 2004, 163–172.
83. S. Ganesan and A. F. Keating, Bisphenol A-induced ovotoxicity involves DNA damage induction to which the ovary mounts a protective response indicated by increased expression of proteins involved in DNA repair and xenobiotic biotransformation, *Toxicol. Sci*, 2016, 169–180.
84. J. Guo, M. H. Zhao and K. T. Shin, *et al.*, The possible molecular mechanisms of bisphenol A action on porcine early embryonic development, *Sci. Rep.*, 2017, 1–9.
85. S. G. Lee, J. Y. Kim and J. Y. Chung, *et al.*, Bisphenol a exposure during adulthood causes augmentation of follicular atresia and luteal regression by Decreasing 17β-estradiol synthesis via downregulation of aromatase in rat ovary, *Environ. Health Perspect.*, 2013, 663–669.
86. C. Liu, W. Duan and L. Zhang, *et al.*, Bisphenol A exposure at an environmentally relevant dose induces meiotic abnormalities in adult male rats, *Cell Tissue Res.*, 2014, 223–232.
87. S. Qi, W. Fu and C. Wang, *et al.*, BPA-induced apoptosis of rat Sertoli cells through Fas/FasL and JNKs/p38 MAPK pathways, *Reprod. Toxicol.*, 2014, 108–116.
88. J. D. Meeker, A. M. Calafat and R. Hauser, Urinary bisphenol A concentrations in relation to serum thyroid and reproductive hormone levels in men from an infertility clinic, *Environ. Sci. Technol.*, 2010, 1458–1463.
89. S. Ehrlich, P. L. Williams and S. A. Missmer, *et al.*, Urinary bisphenol A concentrations and early reproductive health outcomes among women undergoing IVF, *Hum. Reprod.*, 2012, 3583–3592.

90. M. S. Rahman, W. S. Kwon and J. S. Lee, *et al.*, Bisphenol-A Affects Male Fertility via Fertility-related Proteins in Spermatozoa, *Sci. Rep.*, 2015, 1–9.
91. J. D. Meeker, S. Ehrlich and T. L. Toth, *et al.*, Semen quality and sperm DNA damage in relation to urinary bisphenol A among men from an infertility clinic, *Reprod. Toxicol.*, 2010, 532–539.
92. J. Li, R. Mao and Q. Zhou, *et al.*, Exposure to bisphenol A (BPA) in Wistar rats reduces sperm quality with disruption of ERK signal pathway, *Toxicol. Mech. Methods*, 2016, 180–188.
93. P. W. Chu, Z. J. Yang and H. H. Huang, *et al.*, Low-dose bisphenol A activates the ERK signaling pathway and attenuates steroidogenic gene expression in human placental cells†, *Biol. Reprod.*, 2018, 250–258.
94. Q. Zhao, Y. Ma and N. xia Sun, *et al.*, Exposure to bisphenol A at physiological concentrations observed in Chinese children promotes primordial follicle growth through the PI3K/Akt pathway in an ovarian culture system, *Toxicol. In Vitro*, 2014, 1424–1429.
95. Y. Hu, D. Z. Yuan and Y. Wu, *et al.*, Bisphenol A Initiates Excessive Premature Activation of Primordial Follicles in Mouse Ovaries via the PTEN Signaling Pathway, *Reprod. Sci.*, 2018, 609–620.
96. M. Sugiura-Ogasawara, Y. Ozaki, S. I. Sonta, T. Makino and K. Suzumori, Exposure to bisphenol A is associated with recurrent miscarriage, *Hum. Reprod.*, 2005, 2325–2329.
97. W. Tan, H. Huang and Y. Wang, *et al.*, Bisphenol A differentially activates protein kinase C isoforms in murine placental tissue, *Toxicol. Appl. Pharmacol.*, 2013, 163–168.
98. Z. Yang, J. Shi and Z. Guo, *et al.*, A pilot study on polycystic ovarian syndrome caused by neonatal exposure to tributyltin and bisphenol A in rats, *Chemosphere*, 2019, **231**, 151–160.
99. M. Fernández, N. Bourguignon, V. Lux-Lantos and C. Libertun, Neonatal Exposure to Bisphenol A and Reproductive and Endocrine Alterations Resembling the Polycystic Ovarian Syndrome in Adult Rats, *Environ. Health Perspect.*, 2010, **118**(9), 1217–1222.
100. C.-M. Tsen, J.-H. Liu and D.-P. Yang, *et al.*, Study on the correlation of bisphenol A exposure, pro-inflammatory gene expression, and C-reactive protein with potential cardiovascular disease symptoms in young adults, *Environ. Sci. Pollut. Res. Int.*, 2021, 1–2.
101. D. M. Amin, Role of copeptin as a novel biomarker of bisphenol A toxic effects on cardiac tissues: biochemical, histological, immunohistological, and genotoxic study, *Environ. Sci. Pollut. Res.*, 2019, 36037–36047.
102. Y. Hu, L. Zhang and X. Wu, *et al.*, Bisphenol A, an environmental estrogen-like toxic chemical, induces cardiac fibrosis by activating the ERK1/2 pathway, *Toxicol. Lett.*, 2016, 1–9.
103. F. R. Cagampang, C. Torrens, F. W. Anthony and M. A. Hanson, Developmental exposure to bisphenol A leads to cardiometabolic dysfunction in adult mouse offspring, *J. Dev. Origins Health Dis.*, 2012, 287–292.

104. K. Y. Kim, E. Lee and Y. Kim, The Association between Bisphenol A Exposure and Obesity in Children—A Systematic Review with Meta-Analysis, *Int. J. Environ. Res. Public Health*, 2019, **16**, 2521.
105. A. Salehpour, F. Shidfar and M. Hedayati, *et al.*, Bisphenol A enhances adipogenic signaling pathways in human mesenchymal stem cells, *Genes Environ.*, 2020, 13.
106. F. Ahmed, A. Sarsenbayeva, P. Katsogiannos, C. Aguer and M. J. Pereira, The effects of bisphenol A and bisphenol S on adipokine expression and glucose metabolism in human adipose tissue, *Toxicology*, 2020, 152600.
107. J. Bodin, A. K. Bolling and M. Samuelsen, *et al.*, Long-term bisphenol A exposure accelerates insulitis development in diabetes-prone NOD mice, *Immunopharmacol. Immunotoxicol.*, 2013, 349–358.
108. M. K. Silver, M. S. O'Neill, M. F. R. Sowers and S. K. Park, Urinary Bisphenol a and type-2 diabetes in U.S. Adults: Data from NHANES 2003-2008, *PLoS One*, 2011, e26868.
109. Y. Jiang, W. Xia and J. Yang, *et al.*, BPA-induced DNA hypermethylation of the master mitochondrial gene PGC-1α contributes to cardiomyopathy in male rats, *Toxicology*, 2015, **329**, 21–31.
110. T. M. Batista, P. Alonso-Magdalena and E. Vieira, *et al.*, Short-term treatment with Bisphenol-A leads to metabolic abnormalities in adult male mice, *PLoS One*, 2012, e33814.
111. M. S. Bloom, F. S. vom Saal and D. Kim, *et al.*, Serum unconjugated bisphenol A concentrations in men may influence embryo quality indicators during in vitro fertilization, *Environ. Toxicol. Pharmacol.*, 2011, 319–323.
112. Y. Wei, C. Han and S. Li, *et al.*, Maternal exposure to bisphenol A during pregnancy interferes ovaries development of F1 female mice, *Theriogenology*, 2020, 138–148.
113. Y. Ye, Y. Tang, Y. Xiong, L. Feng and X. Li, Bisphenol A exposure alters placentation and causes preeclampsia-like features in pregnant mice involved in reprogramming of DNA methylation of WNT2, *FASEB J.*, 2019, 2732–2742.
114. J. Chevrier, R. B. Gunier and A. Bradman, *et al.*, Maternal urinary bisphenol a during pregnancy and maternal and neonatal thyroid function in the CHAMACOS study, *Environ. Health Perspect.*, 2013, 138–144.
115. J. D. Meeker and K. K. Ferguson, Relationship between urinary phthalate and bisphenol a concentrations and serum thyroid measures in u.s. adults and adolescents from the national health and nutrition examination survey (NHANES) 2007-2008, *Environ. Health Perspect.*, 2011, 1396–1402.
116. Y. F. Zhang, X. M. Ren and Y. Y. Li, *et al.*, Bisphenol A alternatives bisphenol S and bisphenol F interfere with thyroid hormone signaling pathway in vitro and in vivo, *Environ. Pollut.*, 2018, 1072–1079.
117. C. Zhao, P. Xie and T. Yong, *et al.*, MALDI-MS Imaging Reveals Asymmetric Spatial Distribution of Lipid Metabolites from Bisphenol S-Induced Nephrotoxicity, *Anal. Chem.*, 2018, 3196–3204.

118. Y. Shen, T. Liu and Y. Shi, *et al.*, Bisphenol A analogs in patients with chronic kidney disease and dialysis therapy, *Ecotoxicol. Environ. Saf.*, 2019, 109684.
119. W. Peerapanyasut, A. Kobroob, S. Palee, N. Chattipakorn and O. Wongmekiat, *N*-Acetylcysteine Attenuates the Increasing Severity of Distant Organ Liver Dysfunction after Acute Kidney Injury in Rats Exposed to Bisphenol A, *Antioxidants*, 2019, **8**(10), 497.
120. J. M. Braun, K. Yolton and K. N. Dietrich, *et al.*, Prenatal Bisphenol A Exposure and Early Childhood Behavior, *Environ. Health Perspect.*, 2009, **117**(12), 1945–1952.
121. M. Yeo, K. Berglund and M. Hanna, *et al.*, Bisphenol A delays the perinatal chloride shift in cortical neurons by epigenetic effects on the Kcc2 promoter, *Proc. Natl. Acad. Sci. U. S. A.*, 2013, 4315–4320.
122. J. T. Wolstenholme, J. A. Taylor and S. R. J. Shetty, *et al.*, Gestational exposure to low dose bisphenol a alters social behavior in juvenile mice, *PLoS One*, 2011, e25448.
123. S. K. Tiwari, S. Agarwal, A. Tripathi and R. K. Chaturvedi, Bisphenol-A Mediated Inhibition of Hippocampal Neurogenesis Attenuated by Curcumin via Canonical Wnt Pathway, *Mol. Neurobiol.*, 2016, 3010–3029.
124. A. Marmugi, S. Ducheix and F. Lasserre, *et al.*, Low doses of bisphenol a induce gene expression related to lipid synthesis and trigger triglyceride accumulation in adult mouse liver, *Hepatology*, 2012, 395–407.
125. K. Wang, Z. Zhao and W. Ji, Bisphenol A induces apoptosis, oxidative stress and inflammatory response in colon and liver of mice in a mitochondria-dependent manner, *Biomed. Pharmacother.*, 2019, **117**, 109182.
126. L. F. Loffredo, M. E. Coden and S. Berdnikovs, Endocrine Disruptor Bisphenol A (BPA) Triggers Systemic Para-Inflammation and is Sufficient to Induce Airway Allergic Sensitization in Mice, *Nutrients*, 2020, **12**(2), 343.
127. J. Lee and K. T. Lim, Expression of TNF-α and IL-6 in HMC-1 cells treated with bisphenol A is attenuated by plant-originating glycoprotein (75 κda) by blocking p38 MAPK, *Naunyn-Schmiedeberg's Arch. Pharmacol.*, 2010, 51–61.
128. R. Panchanathan, H. Liu, Y. K. Leung, S. mei Ho and D. Choubey, Bisphenol A (BPA) stimulates the interferon signaling and activates the inflammasome activity in myeloid cells, *Mol. Cell. Endocrinol.*, 2015, 45–55.
129. H. Gao, B.-J. Yang and N. Li, *et al.*, Bisphenol A and Hormone-Associated Cancers, *Medicine*, 2015, **94**(1), e211.
130. D. D. Seachrist, K. W. Bonk and S.-M. Ho, *et al.*, A review of the carcinogenic potential of bisphenol A, *Reprod. Toxicol.*, 2016, **59**, 167–182.
131. S. Derouiche, M. Warnier and P. Mariot, *et al.*, Bisphenol A stimulates human prostate cancer cell migration via remodelling of calcium signalling, *SpringerPlus*, 2013, **2**(1), 54.
132. A. Ptak, M. Hoffmann, I. Gruca and J. Barć, Bisphenol A induce ovarian cancer cell migration via the MAPK and PI3K/Akt signalling pathways, *Toxicol. Lett.*, 2014, **229**(2), 357–365.

133. X.-Y. Shi, Z. Wang and L. Liu, *et al.*, Low concentrations of bisphenol A promote human ovarian cancer cell proliferation and glycolysis-based metabolism through the estrogen receptor-α pathway, *Chemosphere*, 2017, **185**, 361–367.
134. K. G. U. Güzel, M. Nazıroğlu and D. Ceyhan, Bisphenol A-Induced Cell Proliferation and Mitochondrial Oxidative Stress Are Diminished via Modulation of TRPV1 Channel in Estrogen Positive Breast Cancer Cell by Selenium Treatment, *Biol. Trace Elem. Res.*, 2020, **198**(1), 118–130.
135. S. Li, B. Wang, Q. Tang, J. Liu and X. Yang, Bisphenol A triggers proliferation and migration of laryngeal squamous cell carcinoma via GPER mediated upregulation of IL-6, *Cell Biochem. Funct.*, 2017, **35**(4), 209–216.
136. B. Lei, S. Sun and X. Zhang, *et al.*, Bisphenol AF exerts estrogenic activity in MCF-7 cells through activation of Erk and PI3K/Akt signals via GPER signaling pathway, *Chemosphere*, 2019, **220**, 362–370.
137. D. Ryszawy, M. Pudełek and P. Kochanowski, *et al.*, High bisphenol A concentrations augment the invasiveness of tumor cells through Snail-1/Cx43/ERRγ-dependent epithelial-mesenchymal transition, *Toxicol. In Vitro*, 2020, **62**, 104676.
138. C.-W. Lin, P.-Y. Lin and P.-C. Yang, Noncoding RNAs in Tumor Epithelial-to-Mesenchymal Transition, *Stem Cells Int.*, 2016, **2016**, 1–13.
139. Y. Zhang and R. A. Weinberg, Epithelial-to-mesenchymal transition in cancer: complexity and opportunities, *Front. Med.*, 2018, **12**(4), 361–373.
140. D. S. Micalizzi, S. M. Farabaugh and H. L. Ford, Epithelial-Mesenchymal Transition in Cancer: Parallels Between Normal Development and Tumor Progression, *J. Mammary Gland Biol. Neoplasia*, 2010, **15**(2), 117–134.
141. C.-H. Heldin, M. Vanlandewijck and A. Moustakas, Regulation of EMT by TGFβ in cancer, *FEBS Lett.*, 2012, **586**(14), 1959–1970.
142. S. Lamouille, J. Xu and R. Derynck, Molecular mechanisms of epithelial–mesenchymal transition, *Nat. Rev. Mol. Cell Biol.*, 2014, **15**(3), 178–196.
143. Y.-S. Kim, K.-A. Hwang and K.-C. Choi, Bisphenol A-induced epithelial-mesenchymal transition was reversed with a phytoestrogen, genistein via oestrogen receptor and downregulation of transforming growth factor-beta signalling pathway, *Endocr. Abstr.*, 2015, 37.
144. B. Huang, N. Luo and X. Wu, *et al.*, The modulatory role of low concentrations of bisphenol A on tamoxifen-induced proliferation and apoptosis in breast cancer cells, *Environ. Sci. Pollut. Res.*, 2019, **26**(3), 2353–2362.
145. G.-A. Lee, K.-A. Hwang and K.-C. Choi, Inhibitory effects of 3,3′-diindolylmethane on epithelial-mesenchymal transition induced by endocrine disrupting chemicals in cellular and xenograft mouse models of breast cancer, *Food Chem. Toxicol.*, 2017, **109**, 284–295.
146. J. Y. Kim, H. G. Choi, H. M. Lee, G. A. Lee, K. A. Hwang and K. C. Choi, Effects of bisphenol compounds on the growth and epithelial

mesenchymal transition of MCF-7 CV human breast cancer cells, *J. Biomed. Res.*, 2017, **31**(4), 358–369.
147. A. Mlynarcikova, L. Macho and M. Fickova, Bisphenol A alone or in combination with estradiol modulates cell cycle- and apoptosis-related proteins and genes in MCF7 cells, *Endocr. Regul.*, 2013, **47**(4), 189–199.
148. K.-A. Hwang, M.-A. Park and N.-H. Kang, *et al.*, Anticancer effect of genistein on BG-1 ovarian cancer growth induced by 17 β-estradiol or bisphenol A via the suppression of the crosstalk between estrogen receptor alpha and insulin-like growth factor-1 receptor signaling pathways, *Toxicol. Appl. Pharmacol.*, 2013, **272**(3), 637–646.
149. C. Fischer, R. Mamillapalli and L. G. Goetz, *et al.*, Bisphenol A (BPA) Exposure In Utero Leads to Immunoregulatory Cytokine Dysregulation in the Mouse Mammary Gland: A Potential Mechanism Programming Breast Cancer Risk, *Horm. Cancer*, 2016, **7**(4), 241–251.
150. C. A. Lamartiniere, S. Jenkins, A. M. Betancourt, J. Wang and J. Russo, Exposure to the endocrine disruptor bisphenol A alters susceptibility for mammary cancer, *Horm. Mol. Biol. Clin. Invest.*, 2011, **5**(2), 45–52.
151. Z. Ma, A. B. Parris and E. W. Howard, *et al.*, In Utero Exposure to Bisphenol a Promotes Mammary Tumor Risk in MMTV-Erbb2 Transgenic Mice Through the Induction of ER-erbB2 Crosstalk, *Int. J. Mol. Sci.*, 2020, **21**(9), 3095.
152. M. A. A. Ibrahim, R. H. Elbakry and N. A. Bayomy, Effect of bisphenol A on morphology, apoptosis and proliferation in the resting mammary gland of the adult albino rat, *Int. J. Exp. Pathol.*, 2016, **97**(1), 27–36.
153. J. Wang, S. Jenkins and C. A. Lamartiniere, Cell proliferation and apoptosis in rat mammary glands following combinational exposure to bisphenol A and genistein, *BMC Cancer*, 2014, **14**(1), 379.
154. P. R. Wadia, N. J. Cabaton and M. D. Borrero, *et al.*, Low-Dose BPA Exposure Alters the Mesenchymal and Epithelial Transcriptomes of the Mouse Fetal Mammary Gland, *PLoS One*, 2013, **8**(5), e63902.
155. G. S. Prins, W.-Y. Hu and G.-B. Shi, *et al.*, Bisphenol A Promotes Human Prostate Stem-Progenitor Cell Self-Renewal and Increases In Vivo Carcinogenesis in Human Prostate Epithelium, *Endocrinology*, 2014, **155**(3), 805–817.
156. Y. Komiya and R. Habas, Wnt signal transduction pathways, *Organogenesis*, 2008, **4**(2), 68–75.
157. C. Gao, Y. Wang and R. Broaddus, *et al.*, Exon 3 mutations of CTNNB1 drive tumorigenesis: a review, *Oncotarget*, 2018, **9**(4), 5492–5508.
158. T. Zhan, N. Rindtorff and M. Boutros, Wnt signaling in cancer, *Oncogene*, 2017, **36**(11), 1461–1473.
159. P. C. Salinas. Wnt Signaling, in *Cellular Migration and Formation of Neuronal Connections*; Elsevier, 2013, pp. 623–638.
160. D. Bastakoty and P. P. Young, Wnt/β-catenin pathway in tissue injury: roles in pathology and therapeutic opportunities for regeneration, *FASEB J.*, 2016, **30**(10), 3271–3284.

161. L. Hui, H. Li and G. Lu, *et al.*, Low Dose of Bisphenol A Modulates Ovarian Cancer Gene Expression Profile and Promotes Epithelial to Mesenchymal Transition Via Canonical Wnt Pathway, *Toxicol. Sci.*, 2018, **164**(2), 527–538.
162. Y. Guo, B. Li and X. Yan, *et al.*, Bisphenol A and polychlorinated biphenyls enhance the cancer stem cell properties of human ovarian cancer cells by activating the WNT signaling pathway, *Chemosphere*, 2020, **246**, 125775.
163. W. Zeng, Bisphenol A triggers the malignancy of nasopharyngeal carcinoma cells via activation of Wnt/β-catenin pathway, *Toxicol. In Vitro*, 2020, **66**, 104881.
164. P. Yaşar, G. Ayaz, S. D. User, G. Güpür and M. Muyan, Molecular mechanism of estrogen-estrogen receptor signaling, *Reprod. Med. Biol.*, 2017, **16**(1), 4–20.
165. N. Fuentes and P. Silveyra, Estrogen receptor signaling mechanisms, *Adv. Protein Chem. Struct. Biol.*, 2019, **116**, 135–170.
166. M. Marino, P. Galluzzo and P. Ascenzi, Estrogen Signaling Multiple Pathways to Impact Gene Transcription, *Curr. Genomics*, 2006, 7(8), 497–508.
167. B. Manavathi and R. Kumar, Steering estrogen signals from the plasma membrane to the nucleus: Two sides of the coin, *J. Cell. Physiol.*, 2006, **207**(3), 594–604.
168. L. Kang, X. Zhang and Y. Xie, *et al.*, Involvement of Estrogen Receptor Variant ER-α36, Not GPR30, in Nongenomic Estrogen Signaling, *Mol. Endocrinol.*, 2010, **24**(4), 709–721.
169. E. R. Prossnitz, J. B. Arterburn and H. O. Smith, *et al.*, Estrogen Signaling through the Transmembrane G Protein–Coupled Receptor GPR30, *Annu. Rev. Physiol.*, 2008, **70**(1), 165–190.
170. A. Ptak and E. L. Gregoraszczuk, Bisphenol A induces leptin receptor expression, creating more binding sites for leptin, and activates the JAK/Stat, MAPK/ERK and PI3K/Akt signalling pathways in human ovarian cancer cell, *Toxicol. Lett.*, 2012, **210**(3), 332–337.
171. H. Song, T. Zhang and P. Yang, *et al.*, Low doses of bisphenol A stimulate the proliferation of breast cancer cells via ERK1/2/ERRγ signals, *Toxicol. In Vitro*, 2015, **30**(1), 521–528.
172. Y. Zhang, F. Wei and J. Zhang, *et al.*, Bisphenol A and estrogen induce proliferation of human thyroid tumor cells via an estrogen-receptor-dependent pathway, *Arch. Biochem. Biophys.*, 2017, **633**, 29–39.
173. S. Dong, S. Terasaka and R. Kiyama, Bisphenol A induces a rapid activation of Erk1/2 through GPR30 in human breast cancer cells, *Environ. Pollut.*, 2011, **159**(1), 212–218.
174. M. Pupo, A. Pisano and R. Lappano, *et al.*, Bisphenol A Induces Gene Expression Changes and Proliferative Effects through GPER in Breast Cancer Cells and Cancer-Associated Fibroblasts, *Environ. Health Perspect.*, 2012, **120**(8), 1177–1182.

175. T. Yaguchi, The endocrine disruptor bisphenol A promotes nuclear ERRγ translocation, facilitating cell proliferation of Grade I endometrial cancer cells via EGF-dependent and EGF-independent pathways, *Mol. Cell. Biochem.*, 2019, **452**(1–2), 41–50.
176. W. Zhang and H. T. Liu, MAPK signal pathways in the regulation of cell proliferation in mammalian cells, *Cell Res.*, 2002, **12**(1), 9–18.
177. M. Katz, I. Amit and Y. Yarden, Regulation of MAPKs by growth factors and receptor tyrosine kinases, *Biochim. Biophys. Acta, Mol. Cell Res.*, 2007, **1773**(8), 1161–1176.
178. Y. Guo, W. Pan and S. Liu, *et al.*, ERK/MAPK signalling pathway and tumorigenesis (Review), *Exp. Ther. Med.*, 2020, **19**(3), 1997–2007.
179. J. Yue and J. M. López, Understanding MAPK Signaling Pathways in Apoptosis, *Int. J. Mol. Sci.*, 2020, **21**(7), 2346.
180. C. A. Taylor, Q. Zheng, Z. Liu and J. E. Thompson, Role of p38 and JNK MAPK signaling pathways and tumor suppressor p53 on induction of apoptosis in response to Ad-eIF5A1 in A549 lung cancer cells, *Mol. Cancer*, 2013, **12**(1), 35.
181. W. Qu, Z. Zhao and S. Chen, *et al.*, Bisphenol A suppresses proliferation and induces apoptosis in colonic epithelial cells through mitochondrial and MAPK/AKT pathways, *Life Sci.*, 2018, **208**, 167–174.
182. H. Song, J. Park and P. T. C. Bui, *et al.*, Bisphenol A induces COX-2 through the mitogen-activated protein kinase pathway and is associated with levels of inflammation-related markers in elderly populations, *Environ. Res.*, 2017, **158**, 490–498.
183. P. Manna and S. K. Jain, Phosphatidylinositol-3,4,5-Triphosphate and Cellular Signaling: Implications for Obesity and Diabetes, *Cell. Physiol. Biochem.*, 2015, **35**(4), 1253–1275.
184. Z. Ding, J. Liang and J. Li, *et al.*, Physical Association of PDK1 with AKT1 Is Sufficient for Pathway Activation Independent of Membrane Localization and Phosphatidylinositol 3 Kinase, *PLoS One*, 2010, **5**(3), e9910.
185. F. Chang, J. T. Lee and P. M. Navolanic, *et al.*, Involvement of PI3K/Akt pathway in cell cycle progression, apoptosis, and neoplastic transformation: a target for cancer chemotherapy, *Leukemia*, 2003, **17**(3), 590–603.
186. N. Xu, Y. Lao, Y. Zhang and D. A. Gillespie, Akt: A Double-Edged Sword in Cell Proliferation and Genome Stability, *J. Oncol.*, 2012, **2012**, 1–15.
187. S. H. Dairkee, M. G. Luciani-Torres, D. H. Moore and W. H. Goodson, Bisphenol-A-induced inactivation of the p53 axis underlying deregulation of proliferation kinetics, and cell death in non-malignant human breast epithelial cells, *Carcinogenesis*, 2013, **34**(3), 703–712.
188. W. H. Goodson, M. G. Luciani and S. A. Sayeed, *et al.*, Activation of the mTOR pathway by low levels of xenoestrogens in breast epithelial cells from high-risk women, *Carcinogenesis*, 2011, **32**(11), 1724–1733.
189. A. Oeckinghaus and S. Ghosh, The NF-κB Family of Transcription Factors and Its Regulation, *Cold Spring Harbor Perspect. Biol.*, 2009, **1**(4), a000034.

190. W. Wang, S. Nag and R. Zhang, Targeting the NFκB Signaling Pathways for Breast Cancer Prevention and Therapy, *Curr. Med. Chem.*, 2014, **22**(2), 264–289.
191. M. Hinz and C. Scheidereit, The IκB kinase complex in NF-κB regulation and beyond, *EMBO Rep.*, 2014, **15**(1), 46–61.
192. A. Israel, The IKK Complex, a Central Regulator of NF-κB Activation, *Cold Spring Harbor Perspect. Biol.*, 2010, **2**(3), a000158.
193. R. Castillo Sanchez, R. Gomez and E. Perez Salazar, Bisphenol A Induces Migration through a GPER-, FAK-, Src-, and ERK2-Dependent Pathway in MDA-MB-231 Breast Cancer Cells, *Chem. Res. Toxicol.*, 2016, **29**(3), 285–295.
194. X.-F. Ma, J. Zhang and H.-L. Shuai, *et al.*, IKKβ/NF-κB mediated the low doses of bisphenol A induced migration of cervical cancer cells, *Arch. Biochem. Biophys.*, 2015, **573**, 52–58.
195. S. Xiong, Y. Wang, H. Li and X. Zhang, Low Dose of Bisphenol A Activates NF-κB/IL-6 Signals to Increase Malignancy of Neuroblastoma Cells, *Cell. Mol. Neurobiol.*, 2017, **37**(6), 1095–1103.
196. G. David, Regulation of oncogene-induced cell cycle exit and senescence by chromatin modifiers, *Cancer Biol. Ther.*, 2012, **13**(11), 992–1000.
197. R. Funayama and F. Ishikawa, Cellular senescence and chromatin structure, *Chromosoma*, 2007, **116**(5), 431–440.
198. S.-M. Ho, R. Rao, S. To, E. Schoch and P. Tarapore, Bisphenol A and its analogues disrupt centrosome cycle and microtubule dynamics in prostate cancer, *Endocr. Relat. Cancer*, 2017, **24**(2), 83–96.
199. A. Bilancio, P. Bontempo and M. Di Donato, *et al.*, Bisphenol A induces cell cycle arrest in primary and prostate cancer cells through EGFR/ERK/p53 signaling pathway activation, *Oncotarget*, 2017, **8**(70), 115620–115631.
200. E. Fatma Karaman, M. Caglayan and S. Sancar-Bas, *et al.*, Global and region-specific post-transcriptional and post-translational modifications of bisphenol A in human prostate cancer cells, *Environ. Pollut.*, 2019, **255**, 113318.
201. S. Elmore, Apoptosis: A Review of Programmed Cell Death, *Toxicol. Pathol.*, 2007, **35**(4), 495–516.
202. D. R. Green and F. Llambi, Cell Death Signaling, *Cold Spring Harbor Perspect. Biol.*, 2015, 7(12), a006080.
203. D. R. McIlwain, T. Berger and T. W. Mak, Caspase Functions in Cell Death and Disease, *Cold Spring Harbor Perspect. Biol.*, 2013, **5**(4), a008656.
204. W. A. Siddiqui, A. Ahad and H. Ahsan, The mystery of BCL2 family: Bcl-2 proteins and apoptosis: an update, *Arch. Toxicol.*, 2015, **89**(3), 289–317.
205. C. Latchoumycandane, S. Vaithinathan, S. C. D'Cruz and P. P. Mathur, Apoptosis and Male Infertility, in *Male Infertility*, Springer International Publishing, Cham, 2020, pp. 479–486.

206. H. Terasaka, Y. Kadoma, H. Sakagami and S. Fujisawa, Cytotoxicity and apoptosis-inducing activity of bisphenol A and hydroquinone in HL-60 cells, *Anticancer Res.*, 2005, **25**(3B), 2241–2247.
207. K. Newton and V. M. Dixit, Signaling in Innate Immunity and Inflammation, *Cold Spring Harbor Perspect. Biol.*, 2012, **4**(3), a006049.
208. W. M. Lewko and R. K. Oldham. Cytokines, in *Principles of Cancer Biotherapy*, Springer Netherlands, Dordrecht, 2009, pp. 155–276.
209. G. Schett and M. F. Neurath, Resolution of chronic inflammatory disease: universal and tissue-specific concepts, *Nat. Commun.*, 2018, **9**(1), 3261.
210. S. Zhang, J. Li, J. Fan and X. Wu, Bisphenol A triggers the malignancy of acute myeloid leukemia cells via regulation of IL-4 and IL-6, *J. Biochem. Mol. Toxicol.*, 2020, **34**(1), e22412.
211. J. B. Spinelli and M. C. Haigis, The multifaceted contributions of mitochondria to cellular metabolism, *Nat. Cell Biol.*, 2018, **20**(7), 745–754.
212. M. D. Brand, A. L. Orr, I. V. Perevoshchikova and C. L. Quinlan, The role of mitochondrial function and cellular bioenergetics in ageing and disease, *Br. J. Dermatol.*, 2013, **169**(2), 1–8.
213. P. H. Reddy, Mitochondrial Medicine for Aging and Neurodegenerative Diseases, *NeuroMol. Med.*, 2008, **10**(4), 291–315.
214. S. H. Kwak, K. S. Park, K.-U. Lee and H. K. Lee, Mitochondrial metabolism and diabetes, *J. Diabetes Invest.*, 2010, **1**(5), 161–169.
215. N. Hammoudi, K. B. Riaz Ahmed, C. Garcia-Prieto and P. Huang, Metabolic alterations in cancer cells and therapeutic implications, *Chin. J. Cancer*, 2011, **30**(8), 508–525.
216. L. Huc, A. Lemarié, F. Guéraud and C. Héliès-Toussaint, Low concentrations of bisphenol A induce lipid accumulation mediated by the production of reactive oxygen species in the mitochondria of HepG2 cells, *Toxicol. In Vitro*, 2012, **26**(5), 709–717.
217. M. K. Moon, M. J. Kim and I. K. Jung, *et al.*, Bisphenol A Impairs Mitochondrial Function in the Liver at Doses below the No Observed Adverse Effect Level, *J. Korean Med. Sci.*, 2012, **27**(6), 644.
218. C. Wang, J. He and T. Xu, *et al.*, Bisphenol A(BPA), BPS and BPB-induced oxidative stress and apoptosis mediated by mitochondria in human neuroblastoma cell lines, *Ecotoxicol. Environ. Saf.*, 2021, **207**, 111299.
219. R. Bagur and G. Hajnóczky, Intracellular Ca^{2+} Sensing: Its Role in Calcium Homeostasis and Signaling, *Mol. Cell*, 2017, **66**(6), 780–788.
220. D. Adiga, R. Radhakrishnan, S. Chakrabarty, P. Kumar and S. P. Kabekkodu, The Role of Calcium Signaling in Regulation of Epithelial-Mesenchymal Transition, *Cells Tissues Organs*, 2020, 1–23.
221. X.-L. Zhang, H.-S. Wang, N. Liu and L.-C. Ge, Bisphenol A stimulates the epithelial mesenchymal transition of estrogen negative breast cancer cells via FOXA1 signals, *Arch. Biochem. Biophys.*, 2015, **585**, 10–16.

222. V. Lloyd, M. Morse and B. Purakal, *et al.*, Hormone-Like Effects of Bisphenol A on p53 and Estrogen Receptor Alpha in Breast Cancer Cells, *Biores. Open Access*, 2019, **8**(1), 169–184.
223. X. Li, W. Xie and C. Xie, *et al.*, Curcumin Modulates miR-19/PTEN/AKT/p53 Axis to Suppress Bisphenol A-induced MCF-7 Breast Cancer Cell Proliferation, *Phyther. Res.*, 2014, **28**(10), 1553–1560.
224. S. Sengupta, I. Obiorah, P. Maximov, R. Curpan and V. Jordan, Molecular mechanism of action of bisphenol and bisphenol A mediated by oestrogen receptor alpha in growth and apoptosis of breast cancer cells, *Br. J. Pharmacol.*, 2013, **169**(1), 167–178.
225. L. P. Jones, A. Sampson and H. J. Kang, *et al.*, Loss of BRCA1 leads to an increased sensitivity to Bisphenol A, *Toxicol. Lett.*, 2010, **199**(3), 261–268.
226. M. Hirao-Suzuki, S. Takeda, K. Okuda, M. Takiguchi and S. Yoshihara, Repeated Exposure to 4-Methyl-2,4-bis(4-hydroxyphenyl)pent-1-ene (MBP), an Active Metabolite of Bisphenol A, Aggressively Stimulates Breast Cancer Cell Growth in an Estrogen Receptor β (ER β)-Dependent Manner, *Mol. Pharmacol.*, 2019, **95**(3), 260–268.
227. E. W. LaPensee, T. R. Tuttle, S. R. Fox and N. Ben-Jonathan, Bisphenol A at Low Nanomolar Doses Confers Chemoresistance in Estrogen Receptor-α–Positive and –Negative Breast Cancer Cells, *Environ. Health Perspect.*, 2009, **117**(2), 175–180.
228. F. Xu, X. Wang and N. Wu, *et al.*, Bisphenol A induces proliferative effects on both breast cancer cells and vascular endothelial cells through a shared GPER-dependent pathway in hypoxia, *Environ. Pollut.*, 2017, **231**, 1609–1620.
229. Y. Tan, Effect of bisphenol A on the EGFR-STAT3 pathway in MCF-7 breast cancer cells, *Mol. Med. Rep.*, 2011, **5**, 41–47.
230. H.-S. Lee, E.-J. Park and J.-H. Oh, *et al.*, Bisphenol A exerts estrogenic effects by modulating CDK1/2 and p38 MAP kinase activity, *Biosci., Biotechnol., Biochem.*, 2014, **78**(8), 1371–1375.
231. M. Li, X. Han, W. Gao, F. Chen and B. Shao, Bisphenol AF stimulates transcription and secretion of C-X-C chemokine ligand 12 to promote proliferation of cultured T47D breast cancer cells, *Toxicology*, 2015, **338**, 30–36.
232. S. Fernandez, Expression and DNA methylation changes in human breast epithelial cells after bisphenol A exposure, *Int. J. Oncol.*, 2012, 369–377.
233. X.-Y. Qin, T. Fukuda and L. Yang, *et al.*, Effects of bisphenol A exposure on the proliferation and senescence of normal human mammary epithelial cells, *Cancer Biol. Ther.*, 2012, **13**(5), 296–306.
234. Y.-S. Kim, K.-A. Hwang and S.-H. Hyun, *et al.*, Bisphenol A and Nonylphenol Have the Potential to Stimulate the Migration of Ovarian Cancer Cells by Inducing Epithelial–Mesenchymal Transition via an Estrogen Receptor Dependent Pathway, *Chem. Res. Toxicol.*, 2015, **28**(4), 662–671.

235. H. Lin, H. Li, G. Lu, Z. Chen, W. Sun, Y. Shi, Z. Fu, B. Huang, X. Zhu, W. Lu and D. Xia, Erratum to: "Low Dose of Bisphenol A. Modulates Ovarian Cancer Gene Expression Profile and Promotes Epithelial to Mesenchymal Transition Via Canonical Wnt Pathway", *Toxicol. Sci*, 2020, **177**(1), 303.
236. A. Ptak, A. Wróbel and E. L. Gregoraszczuk, Effect of bisphenol-A on the expression of selected genes involved in cell cycle and apoptosis in the OVCAR-3 cell line, *Toxicol. Lett.*, 2011, **202**(1), 30–35.
237. A. Ptak and E. L. Gregoraszczuk, Effects of bisphenol A and 17β-estradiol on vascular endothelial growth factor A and its receptor expression in the non-cancer and cancer ovarian cell lines, *Cell Biol. Toxicol.*, 2015, **31**(3), 187–197.
238. M.-A. Park and K.-C. Choi, Effects of 4-Nonylphenol and Bisphenol A on Stimulation of Cell Growth via Disruption of the Transforming Growth Factor-β Signaling Pathway in Ovarian Cancer Models, *Chem. Res. Toxicol.*, 2014, **27**(1), 119–128.
239. A. Ptak, A. Rak-Mardyła and E. L. Gregoraszczuk, Cooperation of bisphenol A and leptin in inhibition of caspase-3 expression and activity in OVCAR-3 ovarian cancer cells, *Toxicol. In Vitro*, 2013, **27**(6), 1937–1943.
240. P. Urriola-Muñoz, R. Lagos-Cabré, D. Patiño-García, J. Reyes and R. Moreno, Bisphenol-A and Nonylphenol Induce Apoptosis in Reproductive Tract Cancer Cell Lines by the Activation of ADAM17, *Int. J. Mol. Sci.*, 2018, **19**(8), 2238.
241. Z.-J. Chen, X.-L. Yang and H. Liu, *et al.*, Bisphenol A modulates colorectal cancer protein profile and promotes the metastasis via induction of epithelial to mesenchymal transitions, *Arch. Toxicol.*, 2015, **89**(8), 1371–1381.
242. K.-S. Zhang, H.-Q. Chen and Y.-S. Chen, *et al.*, Bisphenol A stimulates human lung cancer cell migration via upregulation of matrix metalloproteinases by GPER/EGFR/ERK1/2 signal pathway, *Biomed. Pharmacother.*, 2014, **68**(8), 1037–1043.
243. K.-H. Wang, A.-P. Kao, C.-C. Chang, T.-C. Lin and T.-C. Kuo, Bisphenol A-induced epithelial to mesenchymal transition is mediated by cyclooxygenase-2 up-regulation in human endometrial carcinoma cells, *Reprod. Toxicol.*, 2015, **58**, 229–233.
244. V. C. George and H. P. V. Rupasinghe, DNA damaging and apoptotic potentials of Bisphenol A and Bisphenol S in human bronchial epithelial cells, *Environ. Toxicol. Pharmacol.*, 2018, **60**, 52–57.
245. L. Li, H. Li and J. Zhang, *et al.*, Bisphenol A at a human exposed level can promote epithelial-mesenchymal transition in papillary thyroid carcinoma harbouring BRAF V600E mutation, *J. Cell. Mol. Med.*, 2021, **25**(3), 1739–1749.
246. D. Zhai, J. He, X. Li, L. Gong and Y. Ouyang, Bisphenol A regulates Snail-mediated epithelial-mesenchymal transition in hemangioma cells, *Cell Biochem. Funct.*, 2016, **34**(6), 441–448.

CHAPTER 4

Bisphenol A and Its Impact on Human Telomerase, Telomere Length and Cell Aging

HOAI THI THU TRAN AND EVELYN LAMY*

Molecular Preventive Medicine, University Medical Center and Faculty of Medicine – University of Freiburg, Engesserstraße 4, 79108 Freiburg, Germany
*Email: evelyn.lamy@uniklinik-freiburg.de

4.1 The Human Immune System and Aging

The human immune system plays a crucial role in defense against pathogens such as viruses or mycobacteria and aberrantly differentiating cells, *e.g.* neoplasms, through a plethora of mechanisms. Innate and adaptive immunity mediate these responses with their effector cells. T lymphocytes express a great diversity of T-cell receptors (TCRs) that enable the specific recognition of antigens and the production of memory cells for defense against infection re-occurrence, and help B cells to elicit a proper antibody response. Either an insufficient reaction or overreaction of this T-cell response can affect efficacy against ongoing or latent infection.[1,2] The subpopulation of cytotoxic CD8+ T cells is a critical component of the adaptive immune system. After they have acquired effector function in lymphoid tissues, CD8+ cells circulate in the periphery to "search" for infected or malignant cells. Upon activation, they can secrete high amounts of cytokines, *e.g.* interferon-gamma (IFN-γ) and tumor necrosis factor-alpha (TNF-α), chemokines, perforin, and protease granzyme, which are associated with the anti-viral, anti-tumor, and cytotoxic cell potential.[3] A direct cellular cytolysis and apoptosis induction can be modulated by

Issues in Toxicology No. 43
Bisphenol A: A Multi-modal Endocrine Disruptor
Edited by Natalie R. Gassman

Published by the Royal Society of Chemistry, www.rsc.org

the interaction of the surface death receptor CD95 (also known as Fas) on the target cells and CD95 ligand expressed on CD8+ cells. This mechanism is critical for the maintenance of T-cell homeostasis.[3–5]

As people age, the immune system undergoes marked changes; it becomes slower to respond, and its efficacy to protect against infection and cancer progressively declines. Consequently, the elderly are at greater risk of viral and bacterial infections, reactivation of latent viruses, autoimmune diseases, and neoplasia, as well as the development of age-related disorders.[6,7] Both innate and adaptive immunity undergo age-related alterations, but the latter is postulated as being more affected and will be addressed in more detail in the following. Aging in CD4+ helper T cells has been correlated with impaired responsiveness to pathogens and vaccinations. For instance, a diminished phosphorylation of extracellular signal regulated kinase (ERK) is observed in aged CD4+ cells upon antigen stimulation. This decrease leads to an increase in dual specificity phosphatase 6 (DUSP6) protein phosphorylation and eventually attenuation of miR-181a, an important regulator of T-cell function. In turn, reduction of DUSP6 or restoration of miR-181a level significantly enhanced TCR sensitivity and immune response in T cells from 70–85-year-olds.[8] The CD8+ T-cell response also declines with age, but differently to CD4+ cells.[7] A recent study in human blood provided more insight into the mechanism of age-related impairment of the CD8+ T-cell response. It found that semi-differentiated (known as virtual memory) CD8+ T cells represent 5–10% of the CD8+ cell population in the blood, and exhibit senescence properties, cell proliferation deficiency, and accumulation of DNA damage.[9] Furthermore, a decline in the CD8+ T-cell function in response to the severe acute respiratory syndrome coronavirus 2 (SARS-CoV-2) infection has been correlated with patients' age.[10]

Aging affects the naïve and effector/memory (*i.e.* antigen-exposed) T-cell repertoires differently. The naïve population acquires new T-cell lineages only by emigration from the thymus. Due to the involution of thymic tissue, the production of new naïve T lymphocytes declines during aging with a concomitant accumulation of memory T cells. This could lead to a diminished immune capacity to combat *de novo* antigens, cell regeneration ability, and increased failure to produce memory T-cell responses.[7,11] A reduced number of naïve CD8+ T cells in the circulation is considered a potential biomarker of aging in healthy individuals.[12] This cell population is characterized by a natural loss of CD28 co-stimulatory receptor surface expression with an increased expression of CD57 and programmed cell death protein 1 (PD1). Furthermore, a lower cell proliferation capacity, increased DNA damage frequency with impaired repair capacity, dysregulated mitochondrial function, and shortened telomeres (discussed later) concur with this cell phenotype.[11,13,14] Besides, remodeling of cytokine production towards a pro-inflammatory phenotype with increasing circulation of inflammatory mediators, referred to as "inflamm-aging", has been observed.[11,15] All these parameters have consequently been considered as biomarkers of cell aging and age-associated diseases.

A turnover of the T-cell pool could also be a consequence of excessive replication during persistent exposure to pathogens, *e.g.* cytomegalovirus, or other factors.[7] Thus, it has been proposed that extrinsic factors, including infection, co-mobility, lifestyle, and physical activity contribute to immune aging and regulate cell response with age.[16,17] More evidence is given here for regular physical exercise.[18–22] It helps to reduce pro-inflammatory cytokines, which have been linked to an increased risk of age-associated diseases, and to restore the T-lymphocyte compartment maintaining TCR diversity. An animal study of caloric restriction showed the positive outcomes of maintaining naïve T-cell function and delayed senescence in aged T cells, highlighting a beneficial method to extend lifespan and strengthen the immune function in the elderly.[6,23]

4.2 Telomere Length and Telomerase Regulation in Human T Cells

An effective immune response to antigens depends on lymphocyte proliferation and clonal expansion. This capacity is limited for somatic cells, as first observed in normal fibroblast cultures, due to the inability of conventional DNA polymerases to complete the replication at the 3′ ends of chromosomes, known as the "end replication" problem.[24,25,26] Chromosome ends are capped by telomeres, *i.e.* tandem hexanucleotide (TTAGGG) repeats associated with a complex shelterin protein structure (TRF1, TRF2, RAP1, TIN2, POT1, and TPP1) to maintain genomic stability.[27,28] Approximately 50–200 base pairs of these telomere sequences are then lost with each round of cell division. When telomeres are critically shortened, cells undergo arrest and inevitably senescence. Therefore, telomere length and erosion speed are regarded as the mitotic clock of cell division and aging. The link between aging and shortened telomere repeats has been observed in human cross-sectional and longitudinal studies.[29,30] Additionally, telomere length attrition was reported during human T-cell differentiation as the average telomere length in naïve cells was demonstrated higher than mature, antigen-specific cells.[31–33] Importantly, *in vitro* T-cell long-term cultures revealed that the rate of telomere shortening is not linear. Telomere erosion at the late stage of culture was found to be higher than at earlier time points, indicating maintenance of telomeres in the lymphocyte population.[34,35]

Human telomerase is a ribonucleoprotein enzyme comprising of the telomerase RNA component hTERC, the reverse transcriptase catalytic subunit hTERT, and associated proteins. The canonical function of the telomerase holoenzyme is *de novo* extension of telomeres and helping to maintain cell division capacity.[28] Its high expression was originally reported in the germline, stem cells, and malignant cells,[36] but not in normal somatic cells, which may be due to hTERT silencing. However, it is now clear that telomerase activity is also present in B and T cells, albeit at relatively low levels in the resting state. Robust and transient upregulation of telomerase activity is detected upon mitogen stimulation,[34,37,38] suggesting that telomerase also functions in the human immune response. Increased enzyme activity is also linked with better anti-viral T-cell function by enhancing the production of cytokines, chemokines, and cytotoxic capacity.[14] In contrast, sustained diminished telomerase activity in T cells may lead to cell dysfunction, delayed cellular and humoral immune response, and premature impairment of the immune system.[39,40] Therefore, studies have linked dysregulation of telomeres and telomerase to age-associated diseases, cancer, and early mortality.[41,42]

Telomerase activity gradually decreases with each round of cell activation, along with telomere shortening, as shown in an *in vitro* T-cell long-term culture.[31,43] A recent study also observed a decline in telomerase activity and telomere length in lymphocytes of men with advanced aging.[42] Interestingly, overexpression of hTERT in mice, engineered to be cancer-resistant, delayed telomere erosion, increased the median life span, and ameliorated age-associated pathologies. Such findings suggest a potential anti-aging activity of telomerase in mammalian organisms.[44] Telomerase is tightly regulated in human T cells during cell lineage development and activation. Its activity is intermittent and can be modified by the circadian rhythm, lifestyle, diet, and other factors.[45,46] Estrogen regulates telomerase through hTERT gene expression and promoter activity modulation, as shown in estrogen-targeted cells, or by indirectly targeting the c-Myc transcription factor.[47–49] Furthermore, estradiol can act *via* phosphatidylinositol 3-kinase (PI3K)/protein kinase B (Akt) phosphorylation transduction signaling that controls hTERT posttranscription.[50,51] Some studies observed a reduction in telomerase activity and cell proliferation potential, thereby shortening telomere length, in estrogen-deficient mouse models. These outcomes could be abrogated by estrogen replacement therapy.[52,53]

4.3 Estrogen and Mechanisms of Action in Human Immune Cells

As demonstrated in epidemiologic studies, significant gender-specific differences exist in immune system function. Sex differences occur in innate and adaptive immune responses; sex hormones such as estrogen, progesterone, and androgens all play a part in the differential control of immune responses.[54,55] The presence of putative androgen response elements and estrogen response elements (EREs) in the promoters of several innate immune genes suggests that sex steroids can induce dimorphic innate immune responses.[55,56] It is well known that the effect of estrogen or 17β-estradiol on the immune system will differ based on the concentration of estrogens, cell type, cell activation status, local environment, and experimental conditions.[57] Estrogen has bipotential effects on human monocytes and macrophages, with low (physiological) doses increasing the level of pro-inflammatory cytokines (such as TNF-α and interleukin-6) and high (supraphysiological) doses decreasing the production of these cytokines.[58,59] In the adaptive immune system, all subsets of T cells, including CD4+ (T helper (Th)1, Th2, Th17, and Tregs) and CD8+ cells, are regulated by estrogen.[60–63] For example, at low-level estrogen, Th1 differentiation and cell-mediated immunity were apparent, while at a higher concentration range, the CD4+ T-cell response was switched to Th2.[64] *Via* a negative feedback and or trans-repressive interaction with nuclear transcription factor NF-κB, estrogen controls pro-inflammatory responses in human immune cells.[64,65] Many studies have described estrogen effects and estrogen receptor (ER) signaling on CD8+ and CD4+ T cells.[66] Both T-cell types express ERα and ERβ and are estrogen-sensitive, while higher levels of ERα than ERβ are expressed on circulating female CD4+ T cells, and more receptors are on CD4+ compared to CD8+ T cells.[67,68] Males and females have the same amount of ER expression on CD8+ T cells.[67]

Several downstream signaling cascades have been suggested as estrogen action mechanisms.[69] Crucial here is a fine-tuned balance between ERa and ERb, which function as ligand-activated transcription factors and thus regulate a variety of EREs in estrogen-dependent genes. Ligand binding inside the cytoplasm results in conformational modifications and ER receptor dimerization. These are then transferred to the nucleus and attached to particular ERE or transcription factor complexes (non-ERE) such as c-Fos or c-Jun protein at activator protein 1 binding sites in the regulatory regions of the target genes (known as genomic pathway). In addition, estrogens could function rapidly (within seconds to minutes) by interacting with plasma membrane receptors such as G-protein coupled receptor 30 (GPR30) or membrane ERs, thus activating further downstream signaling cascades and second messenger systems, including intracellular calcium mobilization and phosphorylation (known as the non-genomic pathway).[69]

4.4 Bisphenol A Interference With the Human Immune Response

Bisphenol A (BPA) has been shown to mediate its estrogenic effects at low levels (in the nanomolar range) with a similar or even stronger potency than 17β-estradiol.[70] However, its binding affinity to ERs has been reported to be much weaker,[71] at a concentration of 10^{-6} M, it was almost compatible with that of estrogen at 10^{-8} M for ERα as well as ERβ.[72] Like estrogens, BPA can act as an agonist or antagonist to nuclear and membrane receptors, which leads to the activation of secondary transduction signals. Rapid action of BPA, *i.e.* within minutes, on cell growth through ER-dependent and independent pathways was clearly demonstrated in normal and neoplastic cells. This could then be confirmed in animal models.[70,73–76] Potential adverse effects of "low-dose" BPA have been reported *in vitro*,[77,78] in animal models,[79–81] and in epidemiological studies.[82–85] Increasing evidence points to the potential negative impact of BPA on immunological outcomes,[80,86] but data are still inconsistent, which could be due to the complex mode of action of BPA.[76,87] Exposure of activated peripheral blood mononuclear cells (PBMCs) from healthy donors to 1–10 nM BPA disturbed T-cell function *via* targeting ERs and GPR30. This was evident by a diminished proliferation capacity and altered cytokine production (TNF-α, IFN-γ) upon long-term *in vitro* culture.[78,88] Exposure during development has been proposed to modulate the immune system *via* epigenetic mechanisms, including DNA methylation, histone acetylation, and miRNAs.[87,89] A first cross-sectional study provided a possible link between urinary BPA levels and hypomethylated genes of X-chromosomes related to immune function, inflammation, and metabolism in prepubescent girls.[90] Treatment with BPA at 0.5 $\mu g\,kg^{-1}$ body weight (BW) per day affected the immune system of offspring in rats in terms of transgenerational effects.[80] BPA modulated innate immunity, including phagocytosis activity, by deregulation of effector cell growth, development, and function *in vitro* (10^{-8}–10^{-7} M) and *in vivo* at a dose of 50 $\mu g\,kg^{-1}$ BW per day.[77,91,92]

Animal models demonstrated different CD4+ T-cell polarization patterns towards Th1 and Th2 responses by BPA in a dose- and sex-specific manner.[93–95] Prenatal BPA exposure in mice weakened the immune response and hemostasis in the intestine of the offspring.[79,96] Imbalance of Th1/Th2 and Th17/Treg immunity can cause a marked change in the cytokine profile, compromising immune function, and can promote immunopathology, including autoimmunity and cancers.[97–99] Accordingly, recent publications have proposed a potential link between exposure to endocrine-disrupting chemicals, particularly BPA, and exaggeration of the inflammation of severe acute respiratory syndrome coronavirus 2 (SARS-CoV-2) infection through multiple pathways such as enhanced pro-inflammatory cytokine production and reduced anti-inflammatory cytokine production or directly regulating protein expression, which facilitates virus entry.[100,101] In contrast, a minimal association of BPA exposure and immunological outcomes has been reported in some human studies.[102–107] These studies suggested no correlation between urinary BPA levels and allergies, hay fever, or wheeze, although some observed a significant relationship at a certain time point. For example, the association of prenatal BPA exposure and childhood wheeze was evaluated in 365 mother–infant pairs. Urinary BPA was measured at 16 and 26 weeks of gestation and at birth. BPA was linked to a childhood wheeze at 6 months, but not at 3 years of age.[103] Similarly, in their follow-up study in 5 year-old children, prenatal, but not postnatal BPA exposure was correlated with impairment of lung function and early onset of wheeze.[107] The strengths of these studies are the relatively large sample size with proper quality control and the inclusion of some longitudinal data. However, in these studies, only a single-spot urine sampling was used for BPA analysis, and thus it can't be guaranteed that the timing of the snapshot is representative. Importantly, diet or co-exposure to other factors could have biased the results, even though some covariates had been considered, including socioeconomic and demographic characteristics and genetic background.

4.5 BPA Regulates Telomerase Activity and Telomere Length in Immune Cells

There is limited evidence of BPA exposure affecting telomerase enzyme activity in normal human immune cells. *In vitro* studies demonstrated that "low-dose" exposure to BPA (0.3–3 nM) suppresses telomerase enzyme activity in activated human T cells in a non-monotonic response pattern.[78,108] Then, the non-classical estrogenic pathway was proposed as a relevant mechanism of BPA action on telomerase activity. Using chemical inhibitors, BPA was found to signal through an ERα/ERβ/GPR30-dependent pathway in these cells, which further triggered ERK1/2 transduction signaling. Even though the hTERT promoter contains putative EREs, only minimal changes in hTERT mRNA expression were observed. The inhibition was rather due to other telomerase enzyme activity regulation mechanisms, *i.e.* holoenzyme assembly or nucleus translocation. Telomerase regulation is complex and happens on various levels from transcription to alternative RNA splicing to post-translational modification and translocation of hTERT.[51,109] So, phosphorylation and nuclear translocation of hTERT are involved in activating telomerase enzyme activity in human T lymphocytes.[110] Accordingly, a weaker ability to increase activity in highly differentiated CD8+ cells was proposed as a consequence of changes in Akt-mediated post-translational hTERT phosphorylation.[111] In contrast, increased hTERT mRNA expression was demonstrated in BPA-treated human MCF-7 breast cancer cells at 0.01–1 μM;[112,113] increase in telomerase enzyme activity was then only observed at 1 μM.[113] The discrepancy in the observations between BPA-treated T cells and MCF-7 cells is not surprising, since, depending on the concentration, different effects have already been described, *e.g.* for estrogen regulation of pro- and anti-inflammatory cytokines (see above).[58,59] Besides, there are profound differences in signal regulation between normal and malignant cells, permanent telomerase expression upon malignant transformation being one of them. Also, differences in ER expression between the target cells might be relevant here. As BPA has been suggested to follow a non-monotonic dose–response, it must be kept in mind that effects at higher concentrations cannot be used to predict effects at lower levels.[114] Considering that both events—telomerase inhibition in healthy T cells and further induction in malignant cells—could present adverse effects, this newly described mechanism of BPA should be addressed more closely in the future.

The long-term effects of BPA were further investigated in human CD8+ T cells from healthy male donors.[78] Upon repeated "low-dose" BPA exposure, substantial shortening of average telomere length, along with reduced antiviral IFN-γ cytokine release was observed, which might have been a consequence of BPA attenuation of telomerase activity, cell proliferation, and hTERT protein expression in the cells. These changes suggest that long-term BPA treatment possibly impairs the cytotoxic potential and pushes CD8+ T cells to an aged stage. Of note, BPA did not affect T-cell frequency, and this indicated that telomere erosion was independent of the T-cell repertoire changes from the naïve to the memory cell population. How far these *in vitro* findings can be transferred to humans is currently unknown. In this context, a recent cross-sectional study has demonstrated that higher urinary BPA concentrations were highly correlated with shorter telomere length in peripheral blood cells isolated from Lebanese adult females. However, this study has several limitations, including a small sample size and single-spot urine analysis of BPA.[115] Data correction in the multivariate multinomial regression for possible confounders, including smoking and alcohol consumption, did not change the outcome. These findings are quite preliminary, and larger, longitudinal studies should follow to allow for simultaneous assessment of exposure and outcome to get first evidence of a temporal relationship between the two.

4.6 BPA, DNA Damage and Repair Response in Immune Cells

The genotoxic potential of BPA has been studied in several cell and animal systems, but little information is available from immune cells.[116–120] BPA significantly increased the DNA damage frequency in human PBMCs at 44 nM after 4 h,[116] as determined by comet assay.[116] In another study using the comet assay, the highest tested concentration of 10 nM BPA did not cause DNA damage in activated PBMCs after 24 h exposure. Only in 21 day cultured lymphocytes was significant DNA damage induced and that was already at 1 nM.[108] In these two studies, no further insight into potential underlying mechanisms was given. However, in normal (ERα-negative) non-cancerous breast epithelial cell lines, 72 h exposure to 10 nM BPA was found to upregulate phosphorylation of ataxia-telangiectasia mutated (ATM) and H2AX, both major indicators of DNA double-strand breaks. The transcription factor c-Myc, which is well known to increase proliferation rate and oncogenic transformation,[121,122] was essential for BPA-induced double-strand breaks. Besides damage to the DNA, BPA caused impairment of the DNA repair response in germline cells, cancer cells, and human lymphocytes.[78,123,124] Most of these studies investigated the effect of BPA on gene repair mechanisms at ≥100 μM, and thus will not be addressed further here. A diminished DNA repair capacity in response to exogenous oxidative stress in 24 h BPA (0.3–3 nM)-treated CD8+ T cells has been reported recently. This was independent of intracellular or surface thiol levels and regulation of genes involved in the DNA repair machinery, but associated with a decline in telomerase activity.[78] Human telomerase has been proposed to maintain genomic stability and indirectly enhance DNA damage repair.[125] Cells with ectopic expression of hTERT showed less DNA damage and higher DNA repair kinetics compared to control cells, which might have been due to upregulation of intracellular nucleotides, such as ATP or chromatin-modifying factors.[125] In contrast, inhibition of hTERT caused attenuation of cellular response to ionizing radiation-induced DNA damage *via* blocking of ATM and H2AX phosphorylation, which resulted in premature senescence. Therefore, telomerase has been suggested as an upstream regulator of DNA damage and response.[126]

4.7 Conclusions

In conclusion, extrapolation of so far reported *in vitro* findings to human health is not possible at this time. A deeper understanding of the molecular basis of potential aversive effects might help define to what extent BPA could influence biological outcomes. Consequently, further studies of low-dose BPA on the human immune system, telomerase regulation, aging, and cancer development in animal models and epidemiological studies conducted to the highest possible quality are urgently needed for effect validation.

Abbreviations

TCR	T-cell receptor
IFN-γ	interferon-gamma
TNF-α	tumor necrosis factor-alpha
ERK	extracellular signal regulated kinase
DUSP6	dual specificity phosphatase 6 protein
SARS-CoV-2	severe acute respiratory syndrome coronavirus 2
PD1	programmed cell death protein
hTERC	human telomerase RNA gene
hTERT	human reverse transcriptase catalytic subunit
PI3K	phosphatidylinositol 3-kinase
Akt	protein kinase B
EREs	estrogen response elements
ER	estrogen receptor
Th	T-helper cell
GPR30	G-protein coupled receptor 30
PBMC	peripheral blood mononuclear cell
ATM	ataxia–telangiectasia muted

References

1. P. Corbeau and J. Reynes, Immune reconstitution under antiretroviral therapy: the new challenge in HIV-1 infection, *Blood*, 2011, **117**, 5582–5590.
2. Z. Chen and E. John Wherry, T cell responses in patients with COVID-19, *Nat. Rev. Immunol.*, 2020, **20**, 529–536.
3. P. Wong and E. G. Pamer, CD8 T Cell Responses to Infectious Pathogens, *Annu. Rev. Immunol.*, 2003, **21**, 29–70.
4. P. Bouillet and L. A. O'Reilly, CD95, BIM and T cell homeostasis, *Nat. Rev. Immunol.*, 2009, **9**, 514–519.
5. D. Hassin, O. G. Garber, A. Meiraz, Y. S. Schiffenbauer and G. Berke, Cytotoxic T lymphocyte perforin and Fas ligand working in concert even when Fas ligand lytic action is still not detectable, *Immunology*, 2011, **133**, 190–196.

6. J. Nikolich-Žugich, Ageing and life-long maintenance of T-cell subsets in the face of latent persistent infections, *Nat. Rev. Immunol.*, 2008, **8**, 512–522.
7. J. J. Goronzy and C. M. Weyand, Successful and Maladaptive T Cell Aging, *Immunity*, 2017, **46**, 364–378.
8. G. Li, M. Yu, W. W. Lee, M. Tsang, E. Krishnan, C. M. Weyand and J. J. Goronzy, Decline in miR-181a expression with age impairs T cell receptor sensitivity by increasing DUSP6 activity, *Nat. Med.*, 2012, **18**, 1518–1524.
9. K. M. Quinn, A. Fox, K. L. Harland, B. E. Russ, J. Li and T. H. O. Nguyen, *et al.*, Age-Related Decline in Primary CD8(+) T Cell Responses Is Associated with the Development of Senescence in Virtual Memory CD8(+) T Cells, *Cell Rep.*, 2018, **23**, 3512–3524.
10. J. Westmeier, K. Paniskaki, Z. Karaköse, T. Werner, K. Sutter and S. Dolff, *et al.*, Impaired Cytotoxic CD8+ T Cell Response in Elderly COVID-19 Patients, *mBio*, 2020, **11**, e02243–02220.
11. D. Zhou, M. Borsa and A. K. Simon, Hallmarks and detection techniques of cellular senescence and cellular ageing in immune cells, *Aging Cell*, 2021, **20**, e13316.
12. C. C. Whiting, J. Siebert, A. M. Newman, H. W. Du, A. A. Alizadeh, J. Goronzy, C. M. Weyand, E. Krishnan, C. G. Fathman and H. T. Maecker, Large-Scale and Comprehensive Immune Profiling and Functional Analysis of Normal Human Aging, *PLoS One*, 2015, **10**, e0133627.
13. J. Nikolich-Žugich, The twilight of immunity: emerging concepts in aging of the immune system, *Nat. Immunol.*, 2018, **19**, 10–19.
14. R. B. Effros, Telomere/telomerase dynamics within the human immune system: effect of chronic infection and stress, *Exp. Gerontol.*, 2011, **46**, 135–140.
15. A. Bektas, S. H. Schurman, R. Sen and L. Ferrucci, Human T cell immunosenescence and inflammation in aging, *J. Leukocyte Biol.*, 2017, **102**, 977–988.
16. C. Weyh, K. Krüger and B. Strasser, Physical Activity and Diet Shape the Immune System during Aging, *Nutrients*, 2020, **12**, 622.
17. G. Del Giudice, J. J. Goronzy, B. Grubeck-Loebenstein, P.-H. Lambert, T. Mrkvan, J. J. Stoddard and T. M. Doherty, Fighting against a protean enemy: immunosenescence, vaccines, and healthy aging, *npj Aging Mech. Dis.*, 2017, **4**, 1.
18. N. A. Duggal, G. Niemiro, S. D. R. Harridge, R. J. Simpson and J. M. Lord, Can physical activity ameliorate immunosenescence and thereby reduce age-related multi-morbidity?, *Nat. Rev. Immunol.*, 2019, **19**, 563–572.
19. D. B. Bartlett, L. H. Willis, C. A. Slentz, A. Hoselton, L. Kelly and J. L. Huebner, *et al.*, Ten weeks of high-intensity interval walk training is associated with reduced disease activity and improved innate immune function in older adults with rheumatoid arthritis: a pilot study, *Arthritis Res. Ther.*, 2018, **20**, 127.

20. G. Spielmann, B. K. McFarlin, D. P. O'Connor, P. J. W. Smith, H. Pircher and R. J. Simpson, Aerobic fitness is associated with lower proportions of senescent blood T-cells in man, *Brain, Behav., Immun.*, 2011, **25**, 1521–1529.
21. K. Krüger, K. Alack, R. Ringseis, L. Mink, E. Pfeifer and M. Schinle, *et al.*, Apoptosis of T-Cell Subsets after Acute High-Intensity Interval Exercise, *Med. Sci. Sports Exercise*, 2016, **48**, 2021–2029.
22. M. Gleeson, N. C. Bishop, D. J. Stensel, M. R. Lindley, S. S. Mastana and M. A. Nimmo, The anti-inflammatory effects of exercise: mechanisms and implications for the prevention and treatment of disease, *Nat. Rev. Immunol.*, 2011, **11**, 607–615.
23. I. Messaoudi, J. Warner, M. Fischer, B. Park, B. Hill and J. Mattison, *et al.*, Delay of T cell senescence by caloric restriction in aged long-lived nonhuman primates, *Proc. Natl. Acad. Sci.*, 2006, **103**, 19448–19453.
24. L. Hayflick, The limited in vitro lifetime of human diploid cell strains, *Exp. Cell Res.*, 1965, **37**, 614–636.
25. A. M. Olovnikov, A theory of marginotomy. The incomplete copying of template margin in enzymic synthesis of polynucleotides and biological significance of the phenomenon, *J. Theor. Biol.*, 1973, **41**, 181–190.
26. J. D. Watson, Origin of concatemeric T7 DNA, *Nat.: New biol.*, 1972, **239**, 197–201.
27. E. H. Blackburn, E. S. Epel and J. Lin, Human telomere biology: A contributory and interactive factor in aging, disease risks, and protection, *Science*, 2015, **350**, 1193–1198.
28. C. W. Greider and E. H. Blackburn, Identification of a specific telomere terminal transferase activity in tetrahymena extracts, *Cell*, 1985, **43**, 405–413.
29. S. Ghimire, C. V. Hill, F. S. Sy and R. Rodriguez, Decline in telomere length by age and effect modification by gender, allostatic load and comorbidities in National Health and Nutrition Examination Survey (1999–2002), *PLoS One*, 2019, **14**, e0221690.
30. A. Lustig, H. B. Liu, E. J. Metter, Y. An, M. A. Swaby, P. Elango, L. Ferrucci, R. J. Hodes and N.-P. Weng, Telomere Shortening, Inflammatory Cytokines, and Anti-Cytomegalovirus Antibody Follow Distinct Age-Associated Trajectories in Humans, *Front. Immunol.*, 2017, **8**, 1027.
31. N. P. Weng, B. L. Levine, C. H. June and R. J. Hodes, Human naive and memory T lymphocytes differ in telomeric length and replicative potential, *Proc. Natl. Acad. Sci. U. S. A.*, 1995, **92**, 11091–11094.
32. N. Rufer, T. H. Brummendorf, S. Kolvraa, C. Bischoff, K. Christensen, L. Wadsworth, M. Schulzer and P. M. Lansdorp, Telomere fluorescence measurements in granulocytes and T lymphocyte subsets point to a high turnover of hematopoietic stem cells and memory T cells in early childhood, *J. Exp. Med.*, 1999, **190**, 157–167.

33. A. Roth, H. Yssel, J. Pene, E. A. Chavez, M. Schertzer, P. M. Lansdorp, H. Spits and R. M. Luiten, Telomerase levels control the lifespan of human T lymphocytes, *Blood*, 2003, **102**, 849–857.
34. A. G. Bodnar, N. W. Kim, R. B. Effros and C. P. Chiu, Mechanism of telomerase induction during T cell activation, *Exp. Cell Res.*, 1996, **228**, 58–64.
35. N.-P. Weng, L. D. Palmer, B. L. Levine, H. C. Lane, C. H. June and R. J. Hodes, Tales of tails: regulation of telomere length and telomerase activity during lymphocyte development, differentiation, activation, and aging, *Immunol. Rev.*, 1997, **160**, 43–54.
36. N. W. Kim, M. A. Piatyszek, K. R. Prowse, C. B. Harley, M. D. West, P. L. Ho, G. M. Coviello, W. E. Wright, S. L. Weinrich and J. W. Shay, Specific association of human telomerase activity with immortal cells and cancer, *Science*, 1994, **266**, 2011–2015.
37. N. P. Weng, B. L. Levine, C. H. June and R. J. Hodes, Regulated expression of telomerase activity in human T lymphocyte development and activation, *J. Exp. Med.*, 1996, **183**, 2471–2479.
38. N. Weng, B. L. Levine, C. H. June and R. J. Hodes, Regulation of telomerase RNA template expression in human T lymphocyte development and activation, *J. Immunol.*, 1997, **158**, 3215–3220.
39. B. Molgora, R. Bateman, G. Sweeney, D. Finger, T. Dimler, R. B. Effros and H. F. Valenzuela, Functional assessment of pharmacological telomerase activators in human T cells, *Cells*, 2013, **2**, 57–66.
40. N.-P. Weng, Telomere and adaptive immunity, *Mech. Ageing Dev.*, 2008, **129**, 60–66.
41. G. Aubert and P. M. Lansdorp, Telomeres and aging, *Physiol. Rev.*, 2008, **88**, 557–579.
42. Y. Lin, A. Damjanovic, E. J. Metter, H. Nguyen, T. Truong and K. Najarro, *et al.*, Age-associated telomere attrition of lymphocytes in vivo is co-ordinated with changes in telomerase activity, composition of lymphocyte subsets and health conditions, *Clin. Sci.*, 2015, **128**, 367–377.
43. R. B. Effros, Loss of CD28 expression on T lymphocytes: a marker of replicative senescence, *Dev. Comp. Immunol.*, 1997, **21**, 471–478.
44. A. Tomás-Loba, I. Flores, P. J. Fernández-Marcos, M. L. Cayuela, A. Maraver and A. Tejera, *et al.*, Telomerase reverse transcriptase delays aging in cancer-resistant mice, *Cell*, 2008, **135**, 609–622.
45. W. D. Chen, M. S. Wen, S. S. Shie, Y. L. Lo, H. T. Wo, C. C. Wang, I. C. Hsieh, T. H. Lee and C. Y. Wang, The circadian rhythm controls telomeres and telomerase activity, *Biochem. Biophys. Res. Commun.*, 2014, **451**, 408–414.
46. H. T. T. Tran, N. Schlotz, M. Schreiner and E. Lamy, Short-Term Dietary Intervention with Cooked but Not Raw Brassica Leafy Vegetables Increases Telomerase Activity in CD8+ Lymphocytes in a Randomized Human Trial, *Nutrients*, 2019, **11**, 786.

47. S. Kyo, M. Takakura, T. Kanaya, W. Zhuo, K. Fujimoto, Y. Nishio, A. Orimo and M. Inoue, Estrogen activates telomerase, *Cancer Res.*, 1999, **59**, 5917–5921.
48. S. L. Bayne, M. Jones, H. Li and J.-P. Liu, Estrogen regulation of telomerase in the adrenal gland of mice, *FASEB J.*, 2007, **21**, A1037.
49. S. Nanni, M. Narducci, L. Della Pietra, F. Moretti, A. Grasselli, P. De Carli, A. Sacchi, A. Pontecorvi and A. Farsetti, Signaling through estrogen receptors modulates telomerase activity in human prostate cancer, *J. Clin. Investig.*, 2002, **110**, 219–227.
50. A. B. Engin and A. Engin, The effect of environmental Bisphenol A exposure on breast cancer associated with obesity, *Environ. Toxicol. Pharmacol.*, 2021, **81**, 103544.
51. E. Lamy, V. Goetz, M. Erlacher, C. Herz and V. Mersch-Sundermann, hTERT: another brick in the wall of cancer cells, *Mutat. Res.*, 2013, **752**, 119–128.
52. S. Bayne, M. E. E. Jones, H. Li, A. R. Pinto, E. R. Simpson and J.-P. Liu, Estrogen deficiency leads to telomerase inhibition, telomere shortening and reduced cell proliferation in the adrenal gland of mice, *Cell Res.*, 2008, **18**, 1141–1150.
53. S. Bayne, H. Li, M. E. Jones, A. R. Pinto, M. van Sinderen, A. Drummond, E. R. Simpson and J. P. Liu, Estrogen deficiency reversibly induces telomere shortening in mouse granulosa cells and ovarian aging in vivo, *Protein Cell*, 2011, **2**, 333–346.
54. V. Taneja, Sex Hormones Determine Immune Response, *Front. Immunol.*, 2018, **9**, 1931.
55. S. L. Klein and K. L. Flanagan, Sex differences in immune responses, *Nat. Rev. Immunol.*, 2016, **16**, 626–638.
56. M. F. Hannah, V. B. Bajic and S. L. Klein, Sex differences in the recognition of and innate antiviral responses to Seoul virus in Norway rats, *Brain, Behav., Immun.*, 2008, **22**, 503–516.
57. M. Cunningham and G. Gilkeson, Estrogen receptors in immunity and autoimmunity, *Clin. Rev. Allergy Immunol.*, 2011, **40**, 66–73.
58. W. Gilmore, L. P. Weiner and J. Correale, Effect of estradiol on cytokine secretion by proteolipid protein-specific T cell clones isolated from multiple sclerosis patients and normal control subjects, *J. Immunol.*, 1997, **158**, 446–451.
59. A. Bouman, M. J. Heineman and M. M. Faas, Sex hormones and the immune response in humans, *Hum. Reprod. Update*, 2005, **11**, 411–423.
60. K. Lélu, S. Laffont, L. Delpy, P.-E. Paulet, T. Périnat, S. A. Tschanz, L. Pelletier, B. Engelhardt and J.-C. Guéry, Estrogen Receptor α Signaling in T Lymphocytes Is Required for Estradiol-Mediated Inhibition of Th1 and Th17 Cell Differentiation and Protection against Experimental Autoimmune Encephalomyelitis, *J. Immunol.*, 2011, **187**, 2386–2393.

61. H. P. Priyanka, H. C. Krishnan, R. V. Singh, L. Hima and S. ThyagaRajan, Estrogen modulates in vitro T cell responses in a concentration- and receptor-dependent manner: Effects on intracellular molecular targets and antioxidant enzymes, *Mol. Immunol.*, 2013, **56**, 328–339.
62. D. P. Robinson, O. J. Hall, T. L. Nilles, J. H. Bream and S. L. Klein, 17β-estradiol protects females against influenza by recruiting neutrophils and increasing virus-specific CD8 T cell responses in the lungs, *J. Virol.*, 2014, **88**, 4711–4720.
63. E. Karpuzoglu-Sahin, Y. Zhi-Jun, A. Lengi, N. Sriranganathan and S. Ansar Ahmed, Effects of long-term estrogen treatment on IFN-gamma, IL-2 and IL-4 gene expression and protein synthesis in spleen and thymus of normal C57BL/6 mice, *Cytokine*, 2001, **14**, 208–217.
64. R. H. Straub, The Complex Role of Estrogens in Inflammation, *Endocr. Rev.*, 2007, **28**, 521–574.
65. A. J. Murphy, P. M. Guyre and P. A. Pioli, Estradiol suppresses NF-kappa B activation through coordinated regulation of let-7a and miR-125b in primary human macrophages, *J. Immunol.*, 2010, **184**, 5029–5037.
66. M. Pierdominici, A. Maselli, T. Colasanti, A. M. Giammarioli, F. Delunardo, D. Vacirca, M. Sanchez, A. Giovannetti, W. Malorni and E. Ortona, Estrogen receptor profiles in human peripheral blood lymphocytes, *Immunol. Lett.*, 2010, **132**, 79–85.
67. K. L. Phiel, R. A. Henderson, S. J. Adelman and M. M. Elloso, Differential estrogen receptor gene expression in human peripheral blood mononuclear cell populations, *Immunol. Lett.*, 2005, **97**, 107–113.
68. A. Maret, J. D. Coudert, L. Garidou, G. Foucras, P. Gourdy and A. Krust, *et al.*, Estradiol enhances primary antigen-specific CD4 T cell responses and Th1 development in vivo. Essential role of estrogen receptor alpha expression in hematopoietic cells, *Eur. J. Immunol.*, 2003, **33**, 512–521.
69. D. Khan and S. Ansar, Ahmed, The Immune System Is a Natural Target for Estrogen Action: Opposing Effects of Estrogen in Two Prototypical Autoimmune Diseases, *Front. Immunol.*, 2016, **6**, 635.
70. P. Alonso-Magdalena, A. B. Ropero, S. Soriano, M. García-Arévalo, C. Ripoll, E. Fuentes, I. Quesada and Á. Nadal, Bisphenol-A acts as a potent estrogen via non-classical estrogen triggered pathways, *Mol. Cell. Endocrinol.*, 2012, **355**, 201–207.
71. G. G. Kuiper, J. G. Lemmen, B. Carlsson, J. C. Corton, S. H. Safe, P. T. van der Saag, B. van der Burg and J. A. Gustafsson, Interaction of estrogenic chemicals and phytoestrogens with estrogen receptor beta, *Endocrinology*, 1998, **139**, 4252–4263.
72. H. Hiroi, O. Tsutsumi, M. Momoeda, Y. Takai, Y. Osuga and Y. Taketani, Differential interactions of bisphenol A and 17beta-estradiol with estrogen receptor alpha (ERalpha) and ERbeta, *Endocr. J.*, 1999, **46**, 773–778.

73. S. Dong, S. Terasaka and R. Kiyama, Bisphenol A induces a rapid activation of Erk1/2 through GPR30 in human breast cancer cells, *Environ. Pollut.*, 2011, **159**, 212–218.
74. A. Zsarnovszky, H. H. Le, H.-S. Wang and S. M. Belcher, Ontogeny of Rapid Estrogen-Mediated Extracellular Signal-Regulated Kinase Signaling in the Rat Cerebellar Cortex: Potent Nongenomic Agonist and Endocrine Disrupting Activity of the Xenoestrogen Bisphenol A, *Endocrinology*, 2005, **146**, 5388–5396.
75. A. C. Fitzgerald, C. Peyton, J. Dong and P. Thomas, Bisphenol A and Related Alkylphenols Exert Nongenomic Estrogenic Actions Through a G Protein-Coupled Estrogen Receptor 1 (Gper)/Epidermal Growth Factor Receptor (Egfr) Pathway to Inhibit Meiotic Maturation of Zebrafish Oocytes1, *Biol. Reprod.*, 2015, **93**, 135.
76. I. Cimmino, F. Fiory, G. Perruolo, C. Miele, F. Beguinot, P. Formisano and F. Oriente, Potential Mechanisms of Bisphenol A (BPA) Contributing to Human Disease, *Int. J. Mol. Sci.*, 2020, **21**, 5761.
77. W. Ratajczak-Wrona, M. Rusak, K. Nowak, M. Dabrowska, P. Radziwon and E. Jablonska, Effect of bisphenol A on human neutrophils immunophenotype, *Sci. Rep.*, 2020, **10**, 3083.
78. H. T. T. Tran, C. Herz and E. Lamy, Long-term exposure to "low-dose" bisphenol A decreases mitochondrial DNA copy number, and accelerates telomere shortening in human CD8+ T cells, *Sci. Rep.*, 2020, **10**, 15786.
79. Y. Malaise, S. Menard, C. Cartier, C. Lencina, C. Sommer, E. Gaultier, E. Houdeau and L. Guzylack-Piriou, Consequences of bisphenol a perinatal exposure on immune responses and gut barrier function in mice, *Arch. Toxicol.*, 2018, **92**, 347–358.
80. J. Li, A. Bach, R. B. Crawford, A. S. Phadnis-Moghe, W. Chen and S. D'Ingillo, *et al.*, CLARITY-BPA: Effects of chronic bisphenol A exposure on the immune system: Part 2 – Characterization of lymphoproliferative and immune effector responses by splenic leukocytes, *Toxicology*, 2018, **396–397**, 54–67.
81. S. Santangeli, C. Consales, F. Pacchierotti, H. R. Habibi and O. Carnevali, Transgenerational effects of BPA on female reproduction, *Sci. Total Environ.*, 2019, **685**, 1294–1305.
82. I. A. Lang, T. S. Galloway, A. Scarlett, W. E. Henley, M. Depledge, R. B. Wallace and D. Melzer, Association of urinary bisphenol A concentration with medical disorders and laboratory abnormalities in adults, *Jama*, 2008, **300**, 1303–1310.
83. F. Ranciere, J. Botton, R. Slama, M. Z. Lacroix, L. Debrauwer, M. A. Charles, R. Roussel, B. Balkau and D. J. Magliano, Exposure to Bisphenol A and Bisphenol S and Incident Type 2 Diabetes: A Case-Cohort Study in the French Cohort D.E.S.I.R, *Environ. Health Perspect.*, 2019, **127**, 107013.
84. M. E. Romano, G. M. Webster, A. M. Vuong, R. Thomas Zoeller, A. Chen and A. N. Hoofnagle, *et al.*, Gestational urinary bisphenol A and

maternal and newborn thyroid hormone concentrations: The HOME Study, *Environ. Res.*, 2015, **138**, 453–460.
85. A. Martínez-Ibarra, L. D. Martínez-Razo, K. MacDonald-Ramos, M. Morales-Pacheco, E. R. Vázquez-Martínez, M. López-López, M. Rodríguez Dorantes and M. Cerbón, Multisystemic alterations in humans induced by bisphenol A and phthalates: Experimental, epidemiological and clinical studies reveal the need to change health policies, *Environ. Pollut.*, 2021, **271**, 116380.
86. J. A. Rogers, L. Metz and V. W. Yong, Review: Endocrine disrupting chemicals and immune responses: a focus on bisphenol-A and its potential mechanisms, *Mol. Immunol.*, 2013, **53**, 421–430.
87. J. Xu, G. Huang and T. L. Guo, Developmental Bisphenol A Exposure Modulates Immune-Related Diseases, *Toxics*, 2016, **4**, 23.
88. C. Herz, H. T. T. Tran, S. Landerer, J. Gaus, N. Schlotz and L. Lehr, *et al.*, Normal human immune cells are sensitive to telomerase inhibition by Brassica-derived 3,3-diindolylmethane,partly mediated via ERalpha/beta-AP1 signaling, *Mol. Nutr. Food Res.*, 2017, **61**, 1600524.
89. D. Khan and S. A. Ahmed, Epigenetic Regulation of Non-Lymphoid Cells by Bisphenol A, a Model Endocrine Disrupter: Potential Implications for Immunoregulation, *Front. Endocrinol.*, 2015, **6**, 91.
90. J. H. Kim, L. S. Rozek, A. S. Soliman, M. A. Sartor, A. Hablas, I. A. Seifeldin, J. A. Colacino, C. Weinhouse, M. S. Nahar and D. C. Dolinoy, Bisphenol A-associated epigenomic changes in prepubescent girls: a cross-sectional study in Gharbiah, Egypt, *Environ. Health*, 2013, **12**, 33.
91. A. Roy, S. M. Bauer and B. P. Lawrence, Developmental exposure to bisphenol A modulates innate but not adaptive immune responses to influenza A virus infection, *PLoS One*, 2012, 7, e38448.
92. Y. H. Kuan, F. M. Huang, Y. C. Li and Y. C. Chang, Proinflammatory activation of macrophages by bisphenol A-glycidyl-methacrylate involved NFκB activation via PI3K/Akt pathway, *Food Chem. Toxicol.*, 2012, **50**, 4003–4009.
93. H. Guo, T. Liu, Y. Uemura, S. Jiao, D. Wang and Z. Lin, *et al.*, Bisphenol A in combination with TNF-α selectively induces Th2 cell-promoting dendritic cells in vitro with an estrogen-like activity, *Cell. Mol. Immunol.*, 2010, 7, 227–234.
94. H. Yan, M. Takamoto and K. Sugane, Exposure to Bisphenol A prenatally or in adulthood promotes T(H)2 cytokine production associated with reduction of CD4CD25 regulatory T cells, *Environ. Health Perspect.*, 2008, **116**, 514–519.
95. S. Yoshino, K. Yamaki, R. Yanagisawa, H. Takano, H. Hayashi and Y. Mori, Effects of bisphenol A on antigen-specific antibody production, proliferative responses of lymphoid cells, and TH1 and TH2 immune responses in mice, *Br. J. Pharmacol.*, 2003, **138**, 1271–1276.
96. Y. Malaisé, S. Menard, C. Cartier, E. Gaultier, F. Lasserre and C. Lencina, *et al.*, Gut dysbiosis and impairment of immune system

homeostasis in perinatally-exposed mice to Bisphenol A precede obese phenotype development, *Sci. Rep.*, 2017, 7, 14472.
97. D. S. Robinson, The role of the T cell in asthma, *J. Allergy Clin. Immunol.*, 2010, **126**, 1081–1091, quiz 1092-1083.
98. B. T. Rouse and S. Sehrawat, Immunity and immunopathology to viruses: what decides the outcome?, *Nat. Rev. Immunol.*, 2010, **10**, 514–526.
99. P. Kalinski and M. Moser, Consensual immunity: success-driven development of T-helper-1 and T-helper-2 responses, *Nat. Rev. Immunol.*, 2005, **5**, 251–260.
100. A. Zahra, C. Sisu, E. Silva, S.-C. De Aguiar Greca, H. S. Randeva, K. Chatha, I. Kyrou and E. Karteris, Is There a Link between Bisphenol A (BPA), a Key Endocrine Disruptor, and the Risk for SARS-CoV-2 Infection and Severe COVID-19?, *J. Clin. Med.*, 2020, **9**, 3296.
101. Q. Wu, X. Coumoul, P. Grandjean, R. Barouki and K. Audouze, Endocrine disrupting chemicals and COVID-19 relationships: A computational systems biology approach, *Environ. Int.*, 2020, 106232.
102. E. M. Clayton, M. Todd, J. B. Dowd and A. E. Aiello, The impact of bisphenol A and triclosan on immune parameters in the U.S. population, NHANES 2003-2006, *Environ. Health Perspect.*, 2011, **119**, 390–396.
103. A. J. Spanier, R. S. Kahn, A. R. Kunselman, R. Hornung, Y. Xu, A. M. Calafat and B. P. Lanphear, Prenatal exposure to bisphenol A and child wheeze from birth to 3 years of age, *Environ. Health Perspect.*, 2012, **120**, 916–920.
104. J. H. Savage, E. C. Matsui, R. A. Wood and C. A. Keet, Urinary levels of triclosan and parabens are associated with aeroallergen and food sensitization, *J. Allergy Clin. Immunol.*, 2012, **130**, 453–460, e457.
105. S. V. Vaidya and H. Kulkarni, Association of urinary bisphenol A concentration with allergic asthma: results from the National Health and Nutrition Examination Survey 2005–2006, *J. Asthma*, 2012, **49**, 800–806.
106. K. M. Donohue, R. L. Miller, M. S. Perzanowski, A. C. Just, L. A. Hoepner and S. Arunajadai, *et al.*, Prenatal and postnatal bisphenol A exposure and asthma development among inner-city children, *J. Allergy Clin. Immunol.*, 2013, **131**, 736–742.
107. A. J. Spanier, R. S. Kahn, A. R. Kunselman, E. W. Schaefer, R. Hornung, Y. Xu, A. M. Calafat and B. P. Lanphear, Bisphenol a exposure and the development of wheeze and lung function in children through age 5 years, *JAMA Pediatr*, 2014, **168**, 1131–1137.
108. C. Herz, H. T. T. Tran, N. Schlotz, K. Michels and E. Lamy, Low-dose levels of bisphenol A inhibit telomerase via ER/GPR30-ERK signalling, impair DNA integrity and reduce cell proliferation in primary PBMC, *Sci. Rep.*, 2017, 7, 16631.
109. M. Patrick and N. P. Weng, Expression and regulation of telomerase in human T cell differentiation, activation, aging and diseases, *Cell. Immunol.*, 2019, **345**, 103989.

110. K. Liu, R. J. Hodes and N.-P. Weng, Cutting Edge: Telomerase Activation in Human T Lymphocytes Does Not Require Increase in Telomerase Reverse Transcriptase (hTERT) Protein But Is Associated with hTERT Phosphorylation and Nuclear Translocation, *J. Immunol.*, 2001, **166**, 4826–4830.
111. F. J. Plunkett, O. Franzese, H. M. Finney, J. M. Fletcher, L. L. Belaramani, M. Salmon, I. Dokal, D. Webster, A. D. Lawson and A. N. Akbar, The loss of telomerase activity in highly differentiated CD8+CD28-CD27- T cells is associated with decreased Akt (Ser473) phosphorylation, *J. Immunol.*, 2007, **178**, 7710–7719.
112. A. Takahashi, F. Higashino, M. Aoyagi, S. Kyo, T. Nakata, M. Noda, M. Shindoh, T. Kohgo and H. Sano, Bisphenol A from dental polycarbonate crown upregulates the expression of hTERT, *J. Biomed. Mater. Res., Part B*, 2004, **71**, 214–221.
113. Z. Awada, R. Nasr, R. Akika, A. Ghantous, L. Hou and N. K. Zgheib, Effect of bisphenols on telomerase expression and activity in breast cancer cell lines, *Mol. Biol. Rep.*, 2020, **47**, 3541–3549.
114. L. N. Vandenberg, M. V. Maffini, C. Sonnenschein, B. S. Rubin and A. M. Soto, Bisphenol-A and the great divide: a review of controversies in the field of endocrine disruption, *Endocr. Rev.*, 2009, **30**, 75–95.
115. Z. Awada, F. Sleiman, A. Mailhac, Y. Mouneimne, H. Tamim and N. K. Zgheib, BPA exposure is associated with non-monotonic alteration in ESR1 promoter methylation in peripheral blood of men and shorter relative telomere length in peripheral blood of women, *J. Exposure Sci. Environ. Epidemiol.*, 2019, **29**, 118–128.
116. K. Mokra, A. Kuzminska-Surowaniec, K. Wozniak and J. Michalowicz, Evaluation of DNA-damaging potential of bisphenol A and its selected analogs in human peripheral blood mononuclear cells (in vitro study), *Food Chem. Toxicol.*, 2017, **100**, 62–69.
117. N. Jalal, A. R. Surendranath, J. L. Pathak, S. Yu and C. Y. Chung, Bisphenol A (BPA) the mighty and the mutagenic, *Toxicol. Rep.*, 2017, **5**, 76–84.
118. D. Pfeifer, Y. M. Chung and M. C. Hu, Effects of Low-Dose Bisphenol A on DNA Damage and Proliferation of Breast Cells: The Role of c-Myc, *Environ. Health Perspect.*, 2015, **123**, 1271–1279.
119. E. L. Cavalieri and E. G. Rogan, Is bisphenol A a weak carcinogen like the natural estrogens and diethylstilbestrol?, *IUBMB Life*, 2010, **62**, 746–751.
120. M. Audebert, L. Dolo, E. Perdu, J. P. Cravedi and D. Zalko, Use of the gammaH2AX assay for assessing the genotoxicity of bisphenol A and bisphenol F in human cell lines, *Arch. Toxicol.*, 2011, **85**, 1463–1473.
121. V. H. Cowling and M. D. Cole, The Myc transactivation domain promotes global phosphorylation of the RNA polymerase II carboxy-terminal domain independently of direct DNA binding, *Mol. Cell. Biol.*, 2007, **27**, 2059–2073.

122. E. V. Schmidt, The role of c-myc in cellular growth control, *Oncogene*, 1999, **18**, 2988–2996.
123. P. Allard and M. P. Colaiacovo, Bisphenol A impairs the double-strand break repair machinery in the germline and causes chromosome abnormalities, *Proc. Natl. Acad. Sci. U. S. A.*, 2010, **107**, 20405–20410.
124. Y. Ito, T. Ito, S. Karasawa, T. Enomoto, A. Nashimoto and Y. Hase, *et al.*, Identification of DNA-dependent protein kinase catalytic subunit (DNA-PKcs) as a novel target of bisphenol A, *PLoS One*, 2012, 7, e50481.
125. G. G. Sharma, A. Gupta, H. Wang, H. Scherthan, S. Dhar, V. Gandhi, G. Iliakis, J. W. Shay, C. S. Young and T. K. Pandita, hTERT associates with human telomeres and enhances genomic stability and DNA repair, *Oncogene*, 2003, **22**, 131–146.
126. K. Masutomi, R. Possemato, J. M. Wong, J. L. Currier, Z. Tothova, J. B. Manola, S. Ganesan, P. M. Lansdorp, K. Collins and W. C. Hahn, The telomerase reverse transcriptase regulates chromatin state and DNA damage responses, *Proc. Natl. Acad. Sci. U. S. A.*, 2005, **102**, 8222–8227.

CHAPTER 5

The Role of Bisphenol A in Development of Autoimmunity

ZORA LAZUROVA*[a] AND YEHUDA SHOENFELD[b]

[a] 4th Department of Internal Medicine, Medical Faculty of University of Pavol Jozef Safarik, Rastislavova 43, 040 01 Kosice, Slovakia; [b] Zabludowicz Center for Autoimmune Diseases, Sheba Medical Center, Tel-Aviv, Israel
*Email: zora.lazurova@upjs.sk

5.1 Introduction

There is evidence of a growing prevalence of autoimmune diseases in the past few decades. This can be explained by the development of industrialisation, causing a rising exposure of the population to various environmental factors, including bisphenol A (BPA) and a mixture of other environmental estrogens and xenobiotics.[1,2] BPA as the common endocrine disruptor was first identified to affect the endocrine and mostly reproductive system, but further studies revealed its multisystemic effects.[3] It plays a role in the etiopathogenesis of several cardiovascular, metabolic, neurological, psychiatric, and oncological disorders. Finally, it alters the normal function of the immune system, leading to the development of allergies and autoimmunity.[4–7]

Autoimmunity is defined as a process caused by altered recognition of self-antigens by the immune system, leading to the loss of self-tolerance. The activated immune system then fights against the body's own structures, causing autodamage manifested as autoimmune disease. A wide spectrum of organ-specific and non-specific autoimmune diseases have been described.[8]

Genetic background is generally accepted as a major factor in the etiology of autoimmunity. Since the development of autoimmune disease is usually

Issues in Toxicology No. 43
Bisphenol A: A Multi-modal Endocrine Disruptor
Edited by Natalie R. Gassman

Published by the Royal Society of Chemistry, www.rsc.org

multifactorial, apart from genetics, other exogenous factors are generally included. The classic exogenous risk factors are viral infections, smoking, and hormones, but nowadays the term "mosaic of autoimmunity" is being used. It represents the role of many novel exogenous and environmental factors in the development of autoimmune diseases. Apart from BPA and other endocrine disruptors, environmental factors may create this mosaic, such as geo-epidemiological factors, seasons, ultraviolet light, vitamin D, physical activity, stress, level of hygiene, parasites, certain types of smells, diet, spicy food, coffee, metals, and recently identified adjuvants from vaccines, as well as implanted foreign materials (silicone breast implants, stents and stent-grafts, endoprostheses, *etc.*).[9–16] These factors do not usually induce autoimmune diseases themselves, but they can participate in the development of autoimmune processes in genetically predisposed patients or in combination with multiple risk factors.[17]

Several studies have shown that BPA represents one of the crucial risk factors in the etiopathogenesis of autoimmunity.[1,17–19] Exposure to BPA at any time during the life course may contribute to autoimmune reactions. Moreover, it has been found that prenatal exposure may alter the development of the immune system, induce "pro-autoimmune settings" and predispose the individual to the manifestation of autoimmunity later in adulthood. A wide spectrum of pathophysiological mechanisms is considered, although the data are still limited and sometimes even controversial.

5.2 Pathophysiological Mechanisms of Autoimmunity Due to BPA Exposure

BPA has been shown to affect the immune system by various mechanisms. Not only can exposure to environmental estrogens in adulthood be potentially harmful, but prenatal and early postnatal exposure may also markedly influence the development of the immune system. Moreover, the exposure of the population to a wide spectrum of many other xenoestrogens can also potentiate the effect of BPA and can work synergically in disrupting the normal function of the immune system.[17–19]

BPA itself has a weak estrogenic activity, and it can also affect other hormones involved in the modulation of immune responses. In addition, BPA has been shown to influence various immunological reactions directly. Finally, its exposure can result in the loss of self-tolerance, subsequently leading to autoimmunity.[18,19]

Most effects of BPA are mediated by estrogen receptors (ERs)—ERα and ERβ.[20] Less is known about the orphan receptor estrogen-related receptor-γ (ERRγ, but there is growing evidence about its role in BPA action.[21] Since most immune cells express ERs, BPA may easily regulate their activity, proliferation, or differentiation, with effects similar to estrogen. Apart from typical estrogenic activity, more different ways of action are already known. Due to BPA's structural similarity to steroid and thyroid hormones, it also exerts some affinity to androgen receptors and receptors for thyroid hormones. Several

Table 5.1 Pathophysiological mechanisms of BPA action in the development of autoimmunity.[a]

Type	Site of action		Mechanism
Direct mechanisms	Cellular immunity	Innate immunity	Oxidative stress
			Phagocytosis
			Antigen-presenting cells
			Neutrophil extracellular traps
			Innate lymphoid cells
		Adaptive immunity	Th1/Th2 dominance
			Th17 axis
			Th1-like Th17 cells
			Th22 cells
	Humoral immunity		IgE-mediated autoimmunity
			IgG2 and IgA
Indirect mechanisms	BPA-binding protein		Protein-disulphide isomerase
	Anti-BPA antibodies		Anti-BPA-albumin antibodies
			Anti-BPA-PDI antibodies
	Gut microbiota		Estrogen-like changes
			Local/systemic inflammation
	Hormones		Prolactin
			Thyroid hormones
	Liver function		Elimination of BPA
			Activity of CYP P450
	Epigenetics		Methylation
			Histone modification
			miRNA

[a]Ig: immunoglobulin; PDI: protein disulfide isomerase; Th: T-helper.

other receptors, such as aryl hydrocarbon receptors and peroxisome proliferator-activated receptor (PPARγ), are targets for the BPA molecule.[22–24] Otherwise, few non-specific mechanisms are known to participate in the development of autoimmunity, such as increased oxidative stress, effects on ion channels, and disruption of liver cytochrome activity.[19,25]

BPA may induce the generation of new epitopes by modulation of endogenous molecules and present as a foreign antigen. Furthermore, BPA affects almost all levels of immune responses, from innate immunity, through adaptive reactions, cytokine secretion, and last but not least, the production of antibodies. Therefore, it may also promote or aggravate already pre-existing autoimmune processes.[17] Generally, pathophysiological mechanisms of BPA actions in autoimmunity can be divided into direct and indirect types (Table 5.1).

5.2.1 Direct Mechanisms of BPA Action on the Immune System

5.2.1.1 Cellular Immunity

5.2.1.1.1 Innate Cellular Immunity. The non-specific or innate immune system is the most simple, first line of defense with uniform non-specific responses against infectious and some non-infectious triggers. It presents

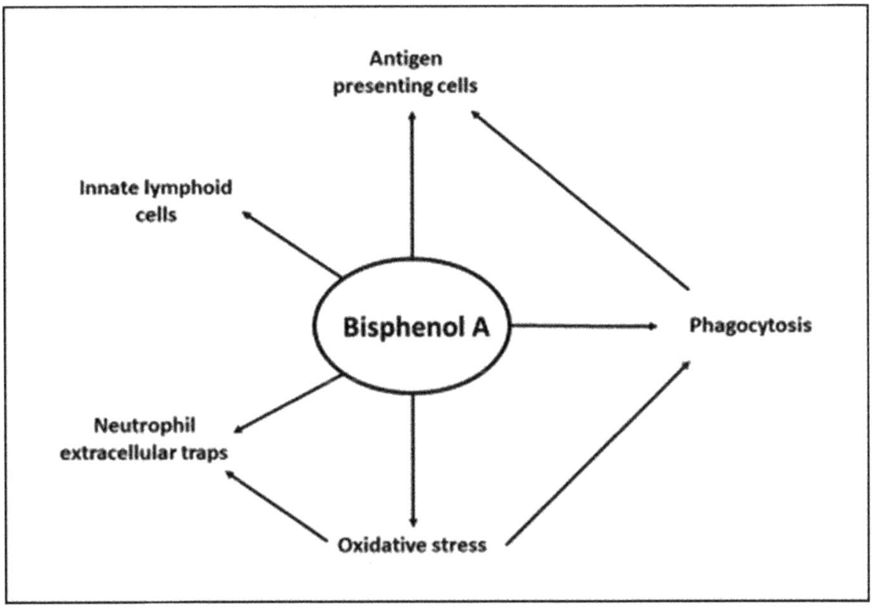

Figure 5.1 The effects of BPA on innate immune responses.

with antigen-independent reactions, not leaving any immunological memory. Innate cellular immunity is represented by phagocytes (mostly macrophages), antigen-presenting cells (APCs) that play a role in the process of phagocytosis, and the activation of naive T-cells, that bride the innate and the adaptive immune responses. Another mechanism of innate defense is oxidative stress and reactive oxygen species (ROS) production that is the least specific and the most primitive form of reaction. Furthermore, newer, less studied forms of innate immunity are neutrophil extracellular traps (NETs) and innate lymphoid cells (ILCs). Alterations in any of these parts of the innate immune system may be implicated in development of autoimmunity. BPA has the potential to impact almost all innate mechanisms[17,18,26,27] (Figure 5.1).

5.2.1.1.1.1 Phagocytosis. Phagocytosis is the major mechanism of the innate immune defense. It represents one of the first-line reactions responsible for eliminating infectious or non-infectious triggers and is considered a bridge to adaptive immunity.

Regulation of the interplay between phagocytic receptors, responsible for initiation and maintenance of mechanisms of phagocytosis, and so-called pattern recognition receptors, such as toll-like receptors (TLRs), which are implicated in distinguishing the self-antigens from "foreign" antigens, are important in the normal function of macrophages. Thus, changes to some of these receptors may increase or decrease phagocytic activity or alter the self-recognition mechanisms that are one of the key factors in self-tolerance breakdown.[26]

Evidence suggests that BPA affects the process of phagocytosis in its various stages, such as chemotaxis and migration of phagocytes and the function of macrophages itself, although the studied results are very inconsistent. BPA has been found to stimulate chemotaxis by activating the CC chemokine ligand (CCL1) signalling pathway.[18,27] In contrast, another study revealed that BPA exposure reduced the chemotactic capacity of neutrophils. TLR-mediated actions of phagocytes have been identified to be stimulated by prenatal BPA exposure in a human study. Overactivated TLRs can bind to autoantigens, thereby contributing to the development of autoimmunity. Stimulation of certain TLRs has lead to changes in cytokine production, too.[28] A study by Yang *et al.* showed that low doses of BPA enhanced the activity of macrophages, but, higher doses induced their apoptosis.[29] Furthermore, it has been identified that BPA has different effects on various subtypes of macrophages (type M1 and M2). BPA, as well as estradiol exposure, promoted the stimulation of M2 migration. These effects have shown to be mediated by either ERα or ERβ, further activating NF-κB (nuclear factor kappa-light-chain-enhancer of activated B cells), JNK (c-Jun N-terminal protein kinase), AP-1 (activator protein 1), or extracellular signal-regulated kinase signaling pathways, known to be responsible for cell stress reactions, controlling transcription of DNA, cytokine production, and cell survival, sometimes generally called the "death" signaling pathways.[30] Another study found that BPA exhibited a synergistic action of ERα and NF-κB transcription factors on NF-kB-associated immune gene expression in macrophages.[31] Dysregulated activation of a proinflammatory mediator NF-κB is associated with chronic inflammatory and autoimmune diseases. It is known to be an important factor in maintaining self-tolerance.[31–33] Moreover, BPA and some other xenoestrogens have been observed to induce interleukin (IL)-10 production in M1 macrophages significantly. In contrast, it decreased IL-10, IL-6, tumour necrosis factor-α (TNFα), and IL-1β release by M2 macrophages.[34]

The process of phagocytosis presents an initial part of physiological and pathological self-destructive responses of the immune system.

5.2.1.1.1.2 Antigen-presenting Cells. The antigen-presenting process is a key mechanism for activation of adaptive immunity, mainly T-cells. APCs, such as dendritic cells or macrophages, present the antigens on their surface after phagocytosis or receptor-mediated endocytosis. Antigens are usually of foreign origin, but autoantigens may be presented too under certain conditions, which leads to autoreactivity in predisposed patients.[35] APCs are crucial in the maintenance of immune tolerance and the regulation of immune responses. Chronic activation of dendritic cells by autoantigens generates auto-reactive T- and B-lymphocytes, causing increased autoantibodies that lead to tissue damage. This mechanism of autoimmunity is common, but typically occurs in the development of systemic lupus erythematosus (SLE).[35,36]

It has been demonstrated that BPA can participate in the development of autoimmunity by altering the antigen-presenting process and by disruption of the elimination of self-activated T-cells, which can play a significant role in etiopathogenesis, as well as in the progression of autoimmune reactions.[17,18]

Dendritic cells exposed to BPA in combination with TNFα promote a CCL1 signalling cascade that induces the T-helper (Th)-2 polarisation preferentially. Overactivity of Th2 lymphocytes and overproduction of their cytokines was previously known as an "autoimmune profile", but more recently, the data about Th1/Th2 dominance in autoimmunity have become more controversial.[17,18,37,38] Most of the published studies are consistent with these data, apart from two studies that describe the induction of Th1 activation. First, Cernadas *et al.* investigated the effect of BPA on the activity of human peripheral blood mononuclear cells and *in vitro* differentiation of dendritic cells from monocytes. BPA has been shown to increase the proliferation of peripheral blood monocytes and up-regulate CD1a expression—a pro-inflammatory biological marker known to trigger Th1 phenotype polarization of naive CD4 T-cells—on dendritic cells.[39]

Another *in vitro* study on cell culture found that BPA exposure significantly increased the expression of ERα on monocyte-derived dendritic cells, while ERβ remained unchanged. BPA promoted the expression of human leukocyte antigen-DR *via* ERα activation, leading to stimulation of pro-inflammatory cytokines. Moreover, the BPA-exposed dendritic cells showed morphological changes as well as altered aggregation.[40]

Altered activity of APCs may play an important role in autoimmune processes by promotion of T-cell abnormalities with a further negative impact on the cytokine production and effects on T-memory and T-suppressor cells well as pathological changes in B-cell function. Therefore, APCs may represent a base for promoting the whole autoimmune cascade.

5.2.1.1.1.3 Innate Lymphoid Cells. ILCs are recently identified hematopoietic cells that regulate lymphocytes and represent a substantial source of cytokine production. Furthermore, ILCs are concerned with innate immune responses against infectious or non-infectious triggers, protection of the epithelial barrier, lymphoid organogenesis and inflammation, tissue remodelling, and regulating the homeostasis of tissue stromal cells. ILCs are morphologically similar to T-cells and B-cells; however, they do not express typical lymphoid differential markers. According to phenotype and function, three classes of ILCs have been recognized (ILC 1, ILC 2, and ILC 3). There is increasing evidence that ILCs play an important role in developing allergies and autoimmunity, mostly the type 3 ILCs, which are proficient in IL-17 production.[41] In addition, ILCs have been identified to participate in the development of multiple sclerosis, inflammatory bowel diseases, psoriasis, and some rheumatic diseases.[42]

Malaisé *et al.* found that prenatal exposure to BPA in mice resulted in decreased ILC 3 in the mucosa of the small intestine, with an increase in immunoglobulin (Ig)G response against commensal bacteria in sera.[43]

Therefore, alteration of ILCs induced by BPA may represent another possible mechanism by which BPA may contribute to the pathogenesis of autoimmunity; however, more studies are needed to elucidate this hypothesis.

5.2.1.1.1.4 Neutrophil Extracellular Traps. NETs are fibrous networks that protrude from the membranes of activated neutrophils. They can be found in various conditions such as infection, malignancy, atherosclerosis, and, last but not least, autoimmune diseases, including rheumatoid arthritis, SLE, anti-neutrophil cytoplasmic antibodies-associated vasculitis, psoriasis, antiphospholipid syndrome, *etc.* Under physiological conditions, NETs are formed as an antimicrobial reaction, although their production may be induced by environmental factors, such as BPA, too.[44–47]

BPA has been found to increase NET production in a dose-dependent manner, although there is still not much data in this field. Tong *et al.* found in their animal *in vivo* research a significant increase in NET production after BPA exposure.[48] In another part of their study, this association was confirmed *in vitro* in human neutrophil cultures, suggesting the stimulating effect of BPA on NET production in humans.[48] However, the exact mechanisms are still unknown, but it can be assumed that ROS represent the bridge between BPA exposure and NET formation, although the direct estrogenic effect on neutrophils cannot be excluded. It is likely that NETs present an alternative mechanism of BPA-induced autoimmunity.

Although NETs represent a physiological mechanism of innate immunity, inadequate promotion of their generation by BPA, leading to the prolonged extracellular exposure to histone complexes, DNA and other autoantigens related to NETs, can play a role in the development of certain systemic autoimmune diseases in some genetically predisposed patients.[49]

5.2.1.1.1.5 Oxidative Stress. ROS play an important role in the development of autoimmunity. Oxidative stress generates damage-associated molecular patterns or autoantigens that further trigger innate and adaptive immunity, leading to dysfunction and cell apoptosis *via* an inflammatory cascade. Cell damage and apoptosis unveil self-antigens, promoting the self-tolerance breakdown in certain predisposed persons and inducing autodestruct reactions. Thus, ROS are factors in the etiopathogenesis of autoimmune diseases generally, but several studies have described their crucial role in developing systemic scleroderma and vitiligo.[50–52]

BPA is known to induce a significant increase in the generation of ROS in a dose-dependent manner. Exposure to even nanomolar doses of BPA causes short, transient bursts of ROS production, but long-term exposure leads to sustained activation of oxidative stress. Free BPA induces ROS production through enzymatic and non-enzymatic mechanisms, generating various free-radical species (superoxides, peroxides, hydroxyl radicals, *etc.*) and cytotoxic and DNA-toxic effects.[51] Additionally, BPA inhibits the expression of antioxidant genes, such as superoxide dismutase, catalase, and the reduced form of glutathione (GSH), contributing to further increases in oxidative stress.[53]

Increased ROS generation by neutrophils has been identified after BPA exposure, which was inhibited by the ERβ antagonists, suggesting that the pro-oxidative effects of BPA are explained by its estrogenic activity.[54] In addition, *in vitro* experimental studies have been confirmed by human clinical trials by Yang *et al.*, who found elevated serum markers of oxidative stress (malondialdehyde and 8-deoxyguanosine) in female postmenopausal patients with increased BPA levels, as well as more recently published data from Wang *et al.*, whose study demonstrated that elevated urinary BPA (and bisphenol F) were associated with significantly higher levels of oxidative stress markers in a population of adult men.[52,55]

ROS production causes DNA damage, oxidation, and fragmentation of intracellular and extracellular proteins. It may also be associated with lipid peroxidation, which all lead to alteration of the cellular organelles and impair normal cellular function. As already mentioned, exposure to BPA is associated with increased levels of 8-deoxyguanosine, which is a marker of DNA damage induced by DNA toxicity of ROS. Similar markers are oxidized DNA base 8-oxoguanine or 8-hydroxy-2′-deoxyguanosine levels that are elevated in certain autoimmune diseases and are considered markers of oxidative DNA damage in the circulating immune complexes in patients with SLE.[52,55,56]

Redox imbalance of membrane lipids or even their peroxidation may lead to compromised functionality and altered structure, which could affect the signal transduction, ion channels, or turn the membrane to be more permeable, which induces cell damage/or programmed apoptosis. ROS may also cause an increase in mitochondrial calcium influx that facilitates mitochondria-dependent apoptosis. Another mechanism found following exposure to phenols is increased expression of transcription factor XBP1 (X-box binding protein 1), an important component of unfolded protein response, and is a significant trigger of the production of immune system mediators.[51,57,58]

Altogether, there is objective evidence that BPA-induced oxidative stress plays a crucial role in the initialisation of processes leading to autoimmunity development. ROS may present a trigger for other mechanisms of innate as well as adaptive immunity.[58]

5.2.1.1.2 Adaptive Cellular Immunity. The adaptive cellular immune responses are mediated by cytokines which are mostly produced and regulated by T-lymphocytes. Cytokines are responsible for the maintenance of interplay between innate cellular mechanisms and immunoglobulin secretion. Therefore, disruption of the cytokine profile may lead to the disbalance of immune function and trigger autoimmunity.[17,18]

Generally, five types of T-cells are known: Th1, Th2, and Th17 cells and the more recently discovered Th1-like Th17 cells and Th22 cells. All types

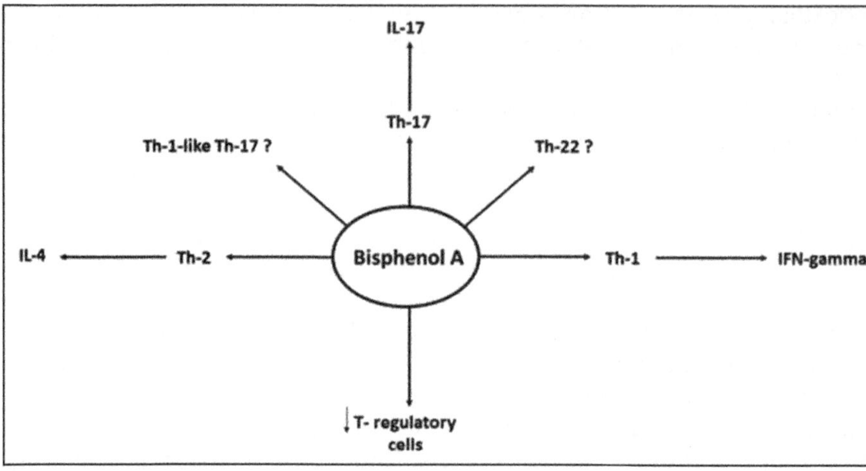

Figure 5.2 BPA effects on T-lymphocytes and cytokine production.

produce pro-inflammatory and anti-inflammatory cytokines with varying predominance.[59]

BPA has been found to impact the activity and differentiation of most types of T-lymphocytes, and therefore to affect cytokine production. The scheme of BPA action on various types of T-lymphocytes is presented in Figure 5.2.

5.2.1.1.2.1 Th1 and Th2 Lymphocytes. Recent data indicate that BPA may induce or contribute to the pathophysiological process of autoimmune diseases by causing an inappropriate polarization of T-cells and shifting of the cytokine profile from Th2 dominance to Th1 dominance, leading to increased activity of the immune system and promoting a breakdown of self-tolerance.[60,61] Th1 cells produce mostly pro-inflammatory cytokines, such as interferon-γ (IFNγ), while Th2 lymphocytes generate predominantly anti-inflammatory cytokines, especially IL-4. It has been found that BPA directly stimulates the production of antigen-specific IFNγ in Th1 cells, but also induces the differentiation and polarization of naive T-cells into the Th1 type.[62] IFNγ itself is a significant stimulator of immune responses; it can promote further pro-inflammatory reactions of innate, adaptive, and humoral immune systems associated with autoimmune reactions.

An *in vivo* animal study demonstrated that BPA in low doses, equivalent to low human exposure, led to the decrease in the number of T-regulatory cells, which promoted the loss of self-tolerance and participated in the switch of T-helper lymphocyte dominance.[63] Moreover, Yoshino *et al.* found that even prenatal exposure to low doses of BPA may lead to upregulation of Th1 responses later in adulthood.[64] Therefore, intrauterine development of the immune system may be altered and programmed to increase autoimmunity development during later stages of life.[64]

BPA affects the cytokine profile by impacting the differentiation of T-cell subtypes, and furthermore, BPA has been observed to affect the secretion of cytokines in all T-lymphocytes directly.[65] Liu *et al.* found a significant increase in pro-inflammatory cytokines—TNFα and IL-6—in contrast with a significant decrease in anti-inflammatory regulatory cytokines, such as IL-10 and transforming growth factor-β, after BPA exposure to a human monocyte-like cell line.[65]

Other studies revealed the stimulatory effects of BPA on Th2 lymphocytes, making the data controversial; the role of Th1/Th2 polarity in autoimmunity remains unclear and seems to differ in various autoimmune diseases.[66–68]

5.2.1.1.2.2 Th17 Lymphocytes. Another possible mechanism by which BPA affects the secretion of cytokines is the alteration of the Th17 axis. BPA has been shown to activate Th17 cells, leading to overproduction of pro-inflammatory cytokine IL-17, inhibiting T-regulatory lymphocyte differentiation and promoting a loss of self-control of the immune system.[69] Furthermore, even prenatal exposure to BPA causes increased proliferation of maternal spleen Th17 lymphocytes and elevated levels of IL-17 in the offspring, which was found to be reversed by vitamin D supplementation during pregnancy.[70]

However, Th17 pathway overactivation against autoantigens causes type 3 immune complex and complement-mediated hypersensitivity, which could play a role in certain autoimmune disorders, such as rheumatoid arthritis. This mechanism has not yet been confirmed in experimental or clinical studies in humans.

5.2.1.2 Humoral Immunity

Production of immunoglobulins against self-antigens is the key step in the development of autoimmune diseases. T-helper lymphocytes mostly regulate antibody secretion by B-cells and plasmatic cells. In addition, it has been found that some endocrine disruptors, including BPA, stimulate a pro-inflammatory cascade of humoral immunity, either directly in the B-cells or by affecting Th-cells and their cytokine secretion.[17,18]

BPA has been identified to cause an increase in IL-4 levels in Th-cells, leading to B-lymphocyte class switching into IgE production, and a direct stimulatory effect of BPA on IgE secretion in plasmatic cells has been found.[71] An experimental study demonstrated elevated urinary IgE levels after BPA exposure.[72] Overproduction of IgE is a significant trigger of inflammatory responses and upregulation of the immune system, leading to allergic reactions and autoimmunity. Previously, it was thought that elevated levels of IgE probably did not induce a self-destructive response, but they were considered to contribute to the already activated autoimmune response. Recently, a special kind, *i.e.* IgE-mediated autoimmunity, has been identified.[73] While its mechanisms have similar patterns to classical allergy induced by exogenous allergens, this IgE-mediated autoimmunity has been termed "autoallergy", and the target molecules are called "autoallergens".[73] Autoantibodies from the IgE class may contribute to the pathogenesis of several autoimmune diseases, such as rheumatoid arthritis, SLE, Graves' disease, multiple sclerosis and certain dermatological autoimmune disorders such as bullous pemphigoid, atopic dermatitis, and psoriasis.[73,74]

Two studies have been published observing the presence of IgE-class antibodies in 32–50% of SLE patients, mostly IgE class antinuclear antibodies (ANAs). In a further subanalysis, IgE class anti-RNP (anti-ribonucleoprotein), anti-dsDNA (anti-double-stranded DNA), anti-SSA/Ro, and anti-Sm (anti-Smith) antibodies have been found too.[75,76] Another study confirmed the relationship between the prevalence of IgE antibodies and the severity of the disease.[77] According to these findings, it is suggested that autoreactive IgE in SLE patients, particularly dsDNA-specific IgE, may be considered a clinical indicator of disease activity. Moreover, disease flare rates were higher in those SLE patients who were positive for circulating IgE-anti-dsDNA antibodies.[76,77] According to Sekigawa *et al.*, the IgE class of anti-SSA/Ro and anti-RNP antibodies are supposed to be responsible for fetal loss in females with SLE.[78]

An IgE subclass of antibodies against a citrullinated peptide in patients with rheumatoid arthritis has been identified too, but their role and clinical importance remain controversial.[79]

Several studies suggest that BPA impacts IgE secretion, and other classes of immunoglobulins may be affected. In the older study of Yurino *et al.*, BPA exposure was identified to stimulate the activity of B1-cells *in vitro*.[80] In a murine model of SLE, the group of BPA-treated mice had stimulated autoantibodies production by B1 cells.[81] An *in vivo* experimental study found increased IgA and IgG2 production after administration of BPA, which were changed due to underlying BPA-induced cytokine disturbances. In addition, this study did not investigate subclasses of IgA, but it is known that increased IgA2 shows significant pro-inflammatory effects. Research in patients with rheumatoid arthritis has found that these patients displayed a shift toward IgA2 subclass production.[82] What is more, BPA may stimulate secretion of immunoglobulin which stimulate autoreactive antibodies in certain conditions, such as pre-existing autoimmunity. Conversely, some data show a possible higher prevalence of autoimmunity in IgA- and IgG2-deficient patients.[83]

There is a lack of data about further detailed pathomechanisms linked to BPA-induced humoral immunity changes; some of them even remain controversial. Nevertheless, according to the aforementioned studies, we can suggest that BPA contributes to autoimmune disease by affecting antibody secretion.

5.2.2 Indirect Mechanisms of BPA Action on Immunity

BPA may influence the function of the immune system by directly affecting immune cells and cytokines, but some other mechanisms indirectly contribute to the development of autoimmunity, for instance by mimicking self structures or presenting as a new epitope, as well as mediation *via* hormonal changes.

5.2.2.1 BPA-binding Protein

It has been found that protein disulfide isomerase (PDI) acts as a binding protein for BPA. Physiologically, PDI is one of the major enzymes in the endoplasmic reticulum that catalyzes the formation of disulfide bonds of nascent or misfolded proteins. In addition, later studies have demonstrated that it also mediates the effects of hormones, especially thyroid hormones and estrogens inside cells, and may act as a hormone reservoir and regulate their intracellular concentrations.[84]

BPA alters the function of the PDI, since it has been found to inhibit isomerase activity of PDI after binding to its molecule; therefore, it may lead to changes in concentrations of some hormones.[85] Moreover, Hiroi *et al.* demonstrated that binding of BPA to PDI causes the formation of complexes which present new unknown epitopes, potentially promoting antigen–antibody immune reactions.[86] PDI is located in a higher volume in the central nervous system; therefore, it probably plays a role in autoimmune disorders predominantly affecting the brain, such as Parkinson's disease, multiple sclerosis, *etc.*[87]

5.2.2.2 Anti-BPA Antibodies

BPA is a foreign molecule in the organism; therefore, it may act as a new antigen, inducing an immune response. In addition, since the structure of BPA may be similar to some endogenous molecules, antibodies against BPA may cross-react with certain self-antigens. Furthermore, as mentioned earlier, BPA can bind to PDI and serum albumin, creating complexes that can represent new epitopes for the immune system.[18]

In studies by Vojdani and co-workers, the presence of antibodies against complexes of BPA-albumin in sera of healthy humans have been detected.[88,89] These autoantibodies can cross-react with some other endogenous structures, such as PDI, myelin basic protein (MBP) and myelin oligodendrocyte protein (MOG). Therefore it can be assumed that antibodies against BPA itself or BPA complexes with endogenous proteins may induce autoimmunity, especially certain neurological autoimmune diseases.[88] Kharazzian *et al.* found a significant association between antibodies against BPA-albumin complexes and the prevalence of α-synuclein antibodies that have been previously identified in patients with Parkinson's disease.[90]

Since the data are poor, it remains unknown whether this effect may be implicated against more endogenous molecules and whether BPA may induce a greater variety of organ-specific or systemic autoimmune disorders by this mechanism.

5.2.2.3 Gut Microbiota

The gut microbiome plays an important role in the development and regulation of the immune system. It is known that variations in gut bacteria may take part in the etiopathogenesis of some cardiovascular or autoimmune diseases. However, there are only limited data about the role of the gut microbiome in participation in the development of autoimmunity. The apoptosis of intestinal epithelial cells due to gut dysbiosis, which enables the presentation of self-antigens inducing the differentiation of autoreactive Th17 cells and other T-helper cells, has been documented in several studies.[91,92]

Xu *et al.* found a shift to a pro-inflammatory profile of gut bacteria after BPA exposure in juvenile mice, which was also associated with an increased risk of diabetes mellitus type 1 in females. BPA exposure was related to increased abundance of *Lachnospiraceae*, *Jeotgalicoccus*, *Ruminococcus*, *Akkermansia*, *Oscillospira*, Rhodospirillales, *Anaerofustis*, and *Turicibacter*, and decreased Sinobacteraceae and Koribacteraceae.[93]

Results of another study also showed significant changes in gut bacteria of mice treated with BPA. They found higher amounts of *Bacteroides*, Mollicutes, Prevotellaceae, Erysipelotrichaceae, *Akkermansia*, *Methanobrevibacter*, and *Sutterella* in individuals exposed to BPA. The same changes in gut microbiome were detected after exposure to ethinylestradiol.[94]

BPA and some other environmental estrogens have been shown to exert estrogenic activity in the gastrointestinal tract, which results in changes in the gut microbiome, possibly altering epithelial function and promoting local and systemic inflammation.[92–94] This is the most probable way BPA affects the gut microbiome and subsequently participates in developing some autoimmune disorders, such as type 1 diabetes mellitus (T1DM), especially Crohn's disease or ulcerative colitis.

5.2.2.4 Hormones

BPA has a molecular structure similar to estrogen; therefore, most BPA effects in organisms are mediated by ERs, which are widely expressed in almost all tissues, including immune cells. Moreover, BPA may mimic some other hormones, such as androgens and thyroid hormones. For example, the molecular mimicry of BPA and thyroid hormones can play a role in developing thyroid autoimmunity, but no further clinical studies have confirmed this hypothesis.[17,18,95,96]

Besides the direct estrogenic activity of BPA, it can also regulate the secretion of several other hormones, such as prolactin, thus indirectly affecting immune responses.

Prolactin is a polypeptide hormone produced in the pituitary gland. Apart from its classic hormonal effects, such as lactation and reproductive functions, it has many other effects, including immune system regulation. There is evidence that prolactin exerts stimulating effects on immunological responses. It affects antigen-presenting functions, promotes responses to any antigens, including self-antigens, increases the secretion of cytokines, and alters B-cell functions, leading to a breakdown of immune self-tolerance.[97,98]

BPA was found to stimulate the production of prolactin, both *in vivo* and *in vitro*.[99] Estrogens are responsible for the increased release of prolactin. Since BPA has agonistic activity on ERs, it also stimulates the production of this hormone in the pituitary gland. Several human studies observed correlations between urinary or serum BPA levels and prolactin.[100,101] Moreover, experimental studies in mice found that BPA exposure significantly promoted the proliferation of pituitary cells and induced prolactin secretion. Furthermore, BPA concentrations positively correlated with the weight of the hypophysis, and higher BPA levels were observed in individuals with prolactinomas compared to a control group.[101]

Several studies have reported an association between increased prolactin levels and autoimmune diseases, such as SLE, antiphospholipid syndrome, rheumatoid arthritis, thyroid autoimmunity, T1DM, adrenal insufficiency, and celiac disease.[97,98] Prolactin may affect all the cellular, adaptive, and humoral immune responses. Both Th1 and Th2 lymphocytes express prolactin receptors, and their activation induces the synthesis of inflammatory cytokines and factors in T-cells, such as IFNγ and IL-2, respectively.[102] In addition, prolactin has been found to regulate the maturation of T-lymphocytes and their differentiation into T-cytotoxic or T-helper phenotypes. The receptors for prolactin are also expressed on the surface of B-cells, thereby directly stimulating autoantibodies in individuals with pre-existing autoimmune diseases. Several studies have demonstrated a role for prolactin in augmenting autoreactivity.[103]

Another study revealed that increased prolactin levels induced a lupus-like phenotype in genetically predisposed individuals. Additionally, an increased number of transgene-expressing B-cells and elevated serum anti-DNA antibody titres were observed in affected animals.[104]

Furthermore, several studies found that hyperprolactinemia can be associated with increased levels of anti-TPO (thyroid peroxidase) and anti-thyroglobulin (TG) antibodies, adrenal antibodies, transglutaminase antibodies, and last but not least, anti-DNA antibodies.[105]

In conclusion, hyperprolactinemia may contribute to the pathogenic processes of BPA-induced autoimmunity by various mechanisms.

5.2.2.5 Liver

Hepatic function may indirectly participate in BPA-induced autoimmunity. BPA is largely eliminated from the organism by the liver. Therefore, impaired liver function disrupts the clearance of BPA. Increased levels of this endocrine disruptor secondarily inhibit CYP450 enzymes, leading to proteomic alterations in immune function.[106,107] Higher concentrations of BPA were associated with the accelerated production of ROS that promote inflammatory responses. Conversely, it is known that patients with liver diseases have a higher prevalence of autoimmune disorders.[108] BPA may be a possible promoter of immune self-reactive responses in these patients, requiring further investigation.

5.2.2.6 Epigenetics

Epigenetics has been observed to have a crucial role in the development of certain autoimmune diseases. BPA was found to induce DNA hyper- or hypo-methylation and histone modification and affection of miRNA synthesis. These changes have the most significant consequences if exposure to the environmental factor is during the prenatal or early postnatal period.[109]

The correlations between increased BPA levels and hypomethylation of the CpG sites of genes responsible for steroidogenesis, immune reactions, and autoimmune thyroid disease have been identified. Whether the BPA-mediated alteration in methylation is related to its estrogenic activity remains unclear. However, a recent study has demonstrated that BPA regulation of methylation was mediated through ERα regulation of DNA methyltransferase (DNMT-1 and -3) expression. Higher urinary BPA levels were found to be associated with hypomethylation of CpG islands on the X-chromosome, attenuating the methylation of genes involved in immune function and transport activity as well as caspase activity.[110]

A previous study by Veiga-Lopez demonstrated that BPA caused an alteration of 45 miRNA out of 742 studied, affecting mostly genes coding enzymes in steroidogenesis.[111] Moreover, BPA was demonstrated to be responsible for hypomethylation in certain genes (Agouti, Pde4d4, Nsbp1, Hoxa) and hypermethylation of Hpcal1 and ER1 and 2.[111] Histone modification is another epigenetic mechanism induced by BPA. Expression of the H3K27me3 gene has been proven to be modified after BPA exposure in humans and mice.[112]

Additionally, BPA has been found to activate estrogen-responsive elements, thereby increasing the expression of certain pro-inflammatory genes, such as NF-κB. Later, BPA-induced epigenetic changes altering expression of more genes implicated in the regulation of immune system function, such as TLRs, have been discovered.[113]

Based on these findings, BPA can promote epigenetic changes affecting genes concerned with immune responses. Unfortunately, few data on clinical relevance in humans have been published.

The epigenetic modifications of DNA expression may be induced by BPA any time during the life course, from early fetal development to late adulthood. Furthermore, exposed individuals may be affected, and transfer to future generations can be possible.

5.3 The Role of BPA in Autoimmune Diseases

Autoimmunity may be targeted against non-specific, systemic antigens, or epitopes located in just one organ: organ-specific autoimmune diseases. BPA has been found to play a role in non-organ-specific as well as organ-specific autoimmunity; however, clinical data in various autoimmune disorders are quite limited and sometimes even controversial.

5.3.1 Systemic (Non-organ-specific) Autoimmune Diseases

5.3.1.1 Systemic Lupus Erythematosus

SLE is an autoimmune disease with a very wide spectrum of clinical manifestations. Apart from genetic background, the major risk factors include viral infections, smoking, and the environment. Among environmental triggers, environmental estrogens, such as BPA, are highly discussed.

In experimental research, several studies demonstrated that BPA induced direct structural damage of DNA that promotes anti-dsDNA autoantibody production.[114] In addition, the generation of NETs caused by BPA are an important mechanism in SLE development. Nevertheless, IgE-mediated autoimmunity has been identified to stimulate antibody production in SLE patients and increase disease activity, as described in Section 5.2.1.2.[75]

Among the most commonly affected organs in patients with SLE are the kidneys. A recently published study in an animal model described a possible role of BPA in exacerbation of pre-existing lupus nephritis, although clinical data on the human population are still missing.[115]

5.3.1.2 Antiphospholipid Syndrome

Antiphospholipid syndrome (APS) is an acquired autoimmune-based thromobophilic state that can occur as either isolated (primary APS) or coexisting with another systemic autoimmune disease, typically with SLE (secondary APS). Unfortunately, there are insufficient data on BPA and its relationship with APS, and only one clinical study has been published.

Sugiura-Ogasawara *et al.* studied a group of women with a history of spontaneous abortion. They demonstrated a significantly higher prevalence of ANA autoantibodies in patients with high serum BPA levels, although the ANA positivity was not associated with abortions themselves. Moreover, this study did not detect any significant correlations between the levels of antiphospholipid antibodies and serum BPA concentrations. According to these data, no relation between BPA and APS has been confirmed. Since it was a small study, further research is needed.[116]

5.3.1.3 Sjögren's Syndrome

Primary Sjögren's syndrome is another systemic autoimmune disease associated with exposure to environmental factors and BPA itself. Although the data on the relationship between the disease and BPA are poor, just one *in vitro* study on dendritic cells from patients suffering from Sjögren's syndrome demonstrated a potentially significant role of BPA in accelerating the disease progression in these patients. After BPA exposure, there were changes in the expression of transcription factors and pro-inflammatory cytokines, leading to further deterioration in this autoimmune disease.[117]

5.3.2 Organ-specific Autoimmune Diseases

5.3.2.1 BPA and Endocrine Autoimmunity

Evidence suggests that BPA, similarly to other xenoestrogens, acts as an endocrine disruptor affecting various endocrine functions, demonstrated in both experimental and large-scale epidemiologic studies.[118,119] BPA has been shown to have the ability to disrupt thyroid function, the hypothalamic–pituitary axis, as well as ovarian and testicular functions, through several mechanisms.[120–123] Although these disrupting mechanisms are relatively well recognized, less is known about the relationship between BPA and endocrine autoimmunity. No data regarding the association of serum or urinary BPA and autoimmune hypophysitis or Addison's disease have been presented to date. Some authors have documented the association with autoimmune thyroid disease, T1DM, and autoimmune ovarian failure.[120–123]

5.3.2.1.1 Autoimmune Thyroid Disease and BPA. While the effects of BPA on thyroid hormone action in experimental and *in vitro* studies seem to be relatively consistent, the association of BPA exposure with thyroid autoimmunity in humans is less recognized. Several studies over the past decade showed that BPA can affect thyroid hormone action through various mechanisms. First, free BPA competitively binds to the thyroid hormone receptor and thereby inhibits gene expression regulated by thyroid hormones.[124] Moreover, this xenoestrogen may also interfere with thyroid hormone synthesis, transport, and partially thyroid hormone metabolism, exerting a potent thyroid-disrupting action.[125]

Epidemiological cross-sectional studies determining the association of serum or urinary BPA concentration with thyroid hormones yielded inconsistent results. Urinary BPA concentrations were negatively associated with total or free T4 in two studies,[126,127] whereas in China, urinary BPA concentrations positively correlated with free T3.[128] A negative relationship between urinary BPA and serum thyroid stimulating hormone levels was observed in studies from Korea and China.[128,129] In contrast, other small-scale studies confirmed the opposite association or did not confirm any association. In addition, two studies assessing urinary BPA concentration in patients with nodular goiter and papillary thyroid carcinoma found significantly higher BPA levels in patients with nodular goiter and those with papillary thyroid cancer than healthy controls.[130,131] Data from the literature on the association of BPA with autoimmune thyroid disease are poor. A recent study from China did not demonstrate any association between urinary BPA concentration and thyroid autoimmunity; similar results were shown in Chinese newborns.[132] Conversely, the Thai 4th National Health Examination Survey reported that serum BPA concentration was independently associated with TPO antibody positivity ($p < 0.001$). The borderline association was also observed with thyroglobulin (TG) antibody positivity ($p < 0.05$).[133] Although others have not confirmed these findings, BPA seems to induce or promote thyroid autoimmunity resulting in thyroid dysfunction. Indirect proof that BPA may induce thyroid autoimmunity is the observation that dental amalgam, which is considered a potential source of human BPA exposure, is associated with anti-TPO and anti-TG positivity. It has been demonstrated that removing mercury-containing dental amalgam has led to a decrease in serum levels of both TPOAb and TGAb.[134] Nevertheless, the mechanisms of action by which BPA alters thyroid autoimmunity require further evaluation.

5.3.2.1.2 BPA and Premature Ovarian Failure. There is evidence that BPA exposure is significantly associated with premature ovarian failure. BPA has been demonstrated to reduce the number of primordial follicles in various animal models, inhibit growth of antral follicles, and disrupts steroid hormone synthesis, resulting in fertility reduction.[135,136] The mechanisms of this ovotoxic effect are still not fully recognized. Whether BPA exposure stimulates the synthesis of antibodies against various ovarian structures is still unknown and requires further clinical studies.

5.3.2.1.3 Diabetes Mellitus Type 1. While the association of BPA exposure to type 2 diabetes mellitus has been confirmed in several studies, some experimental data have suggested a relationship between BPA and T1DM. T1DM is a disease resulting from autoimmune destruction of pancreatic β-cells resulting in insulin deficiency, and secondarily to the T1DM development. Animal studies showed that long-term BPA exposure may accelerate spontaneous insulitis and the development of T1DM,[137,138] especially due to systemic immune alterations.[139] The mechanisms of association of BPA exposure and T1DM development are multifactorial and depend on various factors, including time of exposure, the dose of BPA, sex, age, and state of nutrition.[137] In a study by Xu *et al.*, prenatal BPA exposure had minimal effects on immunity and T1DM development in male offspring. In contrast, it significantly increased the risk of T1DM in female mice.[140]

BPA acts on the pancreatic β-cells by impairing macrophage function, resulting in impaired clearance of apoptotic cells, a common feature of various autoimmune disorders. Endoplasmic reticulum stress after BPA exposure has been demonstrated, and it has led to an experimental model of T1DM in rats.[141] There is only one human study, which has not found any association between urinary BPA concentration and T1DM in children.[142] No other human study about BPA exposure and T1DM development has been performed, and no data on the association of long-term BPA exposure with the presence of autoantibodies specific for autoimmune diabetes mellitus have been published at this time.

5.3.2.2 Autoimmunity in Cardiovascular Diseases

BPA is well known to play a role in the etiopathogenesis of certain non-immunologically-mediated cardiovascular diseases, such as arterial hypertension, coronary artery disease, arrhythmias, and hypertrophic cardiomyopathy.[143–146] However, the data regarding the association of BPA and cardiac autoimmunity are obscure in the current literature.

Myocarditis is a heart disease mostly associated with autoimmune pathogenesis. It can be of infectious and non-infectious origin, although the autoimmune mechanism is also implicated in infectious types. The viral infection by the Coxsackie B3 virus is considered to be the most common cause of myocarditis and can induce an immune reaction against cardiomyocytes. A study by Bruno *et al.* found a relationship between BPA and myocarditis and dilated cardiomyopathy in an experimental animal model. Rats exposed to ordinary human-relevant doses of BPA before viral infection by Coxsackie B3 virus were more prone to develop myocarditis than the control group.[147] Moreover, exposed rats had more severe disease, and they developed dilated cardiomyopathy after myocarditis more commonly.[147] Due to the different distribution and density of ERα and ERβ in males and females, BPA tended to have stronger effects on female myocardium. ERα was found to play a rather protective role, while ERβ signaling participated in myocarditis.[147] BPA has been shown previously to attenuate the expression of ERα in the heart and, conversely, to up-regulate ERβ, contributing to its pro-inflammatory action in the myocardium.[148]

In addition, this study revealed an increased prevalence of pericarditis in affected female rats.[147,149] Evidence suggests that myocarditis may lead to dilated cardiomyopathy due to autoinflammatory damage followed by fibrotic transformation of myocardial tissue. BPA increased the cytokine levels and the intensity of the inflammatory process. Therefore, rats exposed to BPA had a higher prevalence of post-viral-myocarditis dilated cardiomyopathy.[134]

Another study demonstrated that BPA may activate mast cells in the myocardial tissue, largely responsible for increased inflammation activity in females with pre-existing myocarditis, leading to more severe heart damage.[150]

Previously mentioned studies indicate a possible role of BPA in pathogenesis and exacerbation of myocarditis, although all the data have been obtained from animal research. There are no studies about the association of BPA with autoimmune effects on the human heart, which is a research area that needs more focus.

5.3.2.3 Autoimmunity in Gastroenterology

There is a wide spectrum of autoimmune diseases affecting the gastrointestinal, pancreatic, and hepatobiliary tract. As to BPA exposures, the only published data are related to inflammatory bowel disease (IBD). IBD is a group of disorders with two distinct types: Crohn's disease and ulcerative colitis, and other undifferentiated entities. The etiology seems to be multifactorial. Genetic background, environmental factors, and alteration of the gut microbiome may participate in the pathogenesis of IBD. BPA has been shown to exert an estrogenic activity in the colon and thereby alters its function, leading to the development of the local inflammation and the induction of changes in gut microbiota that may influence the local or even systemic immune responses.[92,151] By these mechanisms, BPA probably plays a role in the development of IBD or exacerbation of pre-existing colitis.

An experimental animal study by De Luca *et al.* found an increased clinical and histological activity of colitis in rats exposed to BPA. In addition, higher levels of pro-inflammatory cytokines and changes in concentrations of gut bacterial metabolites have been described, thereby giving a possible explanation to mechanisms of BPA-induced colon inflammation.[152]

Javurek *et al.* demonstrated a significant alteration of gut microbiota after BPA exposure. The changes were almost the same as after ethinylestradiol exposure. Both of them caused sex-dependent changes in gut bacteria colonisation. BPA or synthetic estrogen exposure leads to overgrowth of certain types of bacteria (*i.e. Bacteroides*, Mollicutes, Prevotellaceae, Erysipelotrichaceae, *Akkermansia*, *Methanobrevibacter*, *Sutterella*, *etc.*), that are associated with various diseases, such as IBD, but also with some metabolic disorders and colorectal cancer.[94]

Apart from hyperestrogenism, certain dietary changes belong to the risk factors for IBD development. A diet rich in fat or sugar has been identified to increase the risk of IBD. BPA has been found to induce similar changes in gut microbiome as a high-fat or high-sugar diet. This represents another possible mechanism by which BPA may alter the gut microbiome and thereby contribute to the pathogenesis of autoinflammatory reactions in the bowel.[153]

Several experimental studies found a relationship between BPA and IBD, but there is a lack of data from human clinical research in this field.

The only clinical data available is from an observational National Health and Nutrition Examination Survey study on adults in the United States aged 20–80 years between the years 2005 and 2010, which demonstrated a significant association between exposure phenols and ulcerative colitis. Furthermore, phenols are responsible for chronic diarrheal symptoms. This study observed functional diarrheal symptoms in more than a quarter of responders exposed to phenols, and >42% had abnormal markers of inflammation.[154] However, these data reflected the environmental exposure to phenols in general and were not directly focused on BPA.

In certain gastroenterological autoimmune diseases, such as autoimmune hepatitis, estrogens are a strong risk factor in the etiopathogenesis; therefore, it can be suggested that environmental estrogens with their estrogenic activity may contribute to the development of such disorders. Although, to the best of our knowledge, no studies have yet confirmed such associations.

5.3.2.4 BPA and Autoimmunity in Neurology

The increasing prevalence of certain autoimmune neurological diseases during the past few decades may be associated with exposure to environmental factors, but the data remain controversial. The most studied is the association between BPA and multiple sclerosis.

Rogers *et al.* found a role of gestational BPA exposure in the development of multiple sclerosis in later adulthood in genetically predisposed rats. Furthermore, they demonstrated a sex-specific response with higher inducibility of this autoimmune disease in male individuals.[155] Conversely, the results of an older study byKrementsov *et al.* do not support this data, and the authors do not consider developmental exposure to BPA as a risk factor for multiple sclerosis.[156] Moreover, neither of these studies described a relationship between BPA exposure in adulthood and the higher prevalence of multiple sclerosis.[155,156]

Similarly, as in viral myocarditis, BPA stimulates the inflammation process in Theiler's virus-induced demyelination (TVID), a mouse model of multiple sclerosis. BPA amplified immune-related gene expression in infected mice, increased inflammation, and accelerated the onset of TVID symptoms.[157]

Several experimental animal models have demonstrated that BPA may contribute to the development of autoimmune encephalomyelitis and multiple sclerosis, but very few data have been published about such possible effects in humans.

As described earlier, BPA binds to serum albumin and PDI, and these complexes present as new potential epitopes, inducing antibody production. This may lead to cross-reactivity with MBP and MOG, both key molecules in the development of neuroautoimmunity. Kharrazian and Vojdani found a higher prevalence of anti-MBP and anti-MOG autoantibodies in patients with coexisting autoantibodies against BPA–albumin and BPA–PDI complexes.[158] This mechanism may participate in autoimmune demyelination diseases. Whether the autoantibodies to BPA–albumin complexes play a role in other autoimmune neurological disorders, needs to be elucidated.

Recently, there has been growing evidence about the possible autoimmune background of Parkinson's disease. The data are not consistent, but α-synuclein, being a target for autoimmunity, is highly debated, and specific changes in innate and adaptive immune reactions in patients with this neurodegenerative disease.[159] Kharrazian *et al.* demonstrated that patients with anti-BPA–albumin complex antibodies had a higher prevalence of autoantibodies against α-synuclein and identified a significant positive correlation between the levels of these antibodies.[160] Thereby, BPA may promote the autoimmune processes leading to Parkinson's disease.

5.3.2.5 BPA and ASIA Syndrome

The autoimmune/inflammatory syndrome induced by adjuvants (ASIA syndrome), also known as Shoenfeld's syndrome, represents a group of several immune-mediated conditions that are associated with exposure to substances with adjuvant activity, such as silicones present in various types of implants, *i.e.* breast implants, or adjuvants used in vaccines may lead to a post-vaccination phenomenon. The manifestation of ASIA syndrome is very heterogeneous. Severity may vary from very mild to severe. The clinical symptoms may range from non-specific, such as fatigue, malaise, or myalgia, up to fully presented systemic or organ-specific autoimmune disease. According to their pathophysiological mechanisms of action, three types of adjuvant substances have been recognized: active immune stimulants; carriers; and the last group, vehicle adjuvants, help inoculate the antigens.[161,162]

Although no data about BPA and its possible role as an adjuvant have been published yet, BPA is a common component in various materials used in medicine, such as dental fillings, multiple types of implants, prosthetic materials, *etc.* That leads to the question whether BPA could be considered as an adjuvant and whether BPA can also be one of the triggers inducing ASIA syndrome.

5.4 Conclusion

There is growing evidence that environmental factors, such as BPA, play an important role in the etiopathogenesis of autoimmunity. BPA affects the immune system directly and *via* many indirect mechanisms, leading to potential loss of self-tolerance and autoimmune diseases. Several experimental and observational animal studies, as well as human studies, have found a relationship between BPA and systemic autoimmune diseases, such as SLE and Sjögren's syndrome, as well as organ-specific autoimmunity of thyroid gland, heart, central nervous system, large bowel, and diabetes mellitus. Even prenatal exposure to BPA can lead to the immune system being more prone to developing autoimmunity in later adulthood. However, in many cases, the results are still controversial and inconsistent. Finally, it is thought that BPA participates in the pathogenesis of several autoimmune diseases, but further research is required into the mechanisms of action.

References

1. I. Jochmanová, Z. Lazúrová and M. Rudnay, *et al.*, Environmental estrogen bisphenol A and autoimmunity, *Lupus*, 2015, **24**(4–5), 392–399.
2. B. S. Rubin, Bisphenol A: an endocrine disruptor with widespread exposure and multiple effects, *J. Steroid Biochem. Mol. Biol.*, 2011, **127**(1–2), 27–34.
3. A. Tomza-Marciniak, P. Stępkowska and J. Kuba, *et al.*, Effect of bisphenol A on reproductive processes: A review of in vitro, in vivo and epidemiological studies, *J. Appl. Toxicol.*, 2018, **38**(1), 51–80.
4. J. R. Rochester, Bisphenol A and human health: a review of the literature, *Reprod. Toxicol.*, 2013, **42**, 132–155.
5. Z. Lazurova and I. Lazurova, The environmental estrogen bisphenol A and its effects on the human organism, *Vnitr. Lek.*, 2014, **59**(6), 466–471.
6. X. Gao and H. S. Wang, Impact of bisphenol a on the cardiovascular system - epidemiological and experimental evidence and molecular mechanisms, *Int. J. Environ. Res. Public Health*, 2014, **11**(8), 8399–8413.
7. D. D. Seachrist, K. W. Bonk and S. M. Ho, *et al.*, A review of the carcinogenic potential of bisphenol A, *Reprod. Toxicol.*, 2016, **59**, 167–182.

8. L. Wang, F. S. Wang and M. E. Gershwin, Human autoimmune diseases: a comprehensive update, *J. Intern. Med.*, 2015, **278**(4), 369–395.
9. T. Bashi, G. Bizzaro and D. Ben-Ami Shor, *et al.*, The mechanisms behind helminth's immunomodulation in autoimmunity, *Autoimmun. Rev.*, 2015, **14**(2), 98–104.
10. L. E. Guimarães, B. Baker and C. Perricone, *et al.*, Vaccines, adjuvants and autoimmunity, *Pharmacol. Res.*, 2015, **100**, 190–209.
11. C. Perricone, N. Shoenfeld and N. Agmon-Levin, *et al.*, Smell and autoimmunity: a comprehensive review, *Clin. Rev. Allergy Immunol.*, 2013, **45**(1), 87–96.
12. E. A. Yamamoto and T. N. Jørgensen, Relationships Between Vitamin D, Gut Microbiome, and Systemic Autoimmunity, *Front. Immunol.*, 2020, **10**, 3141.
13. Y. Rosen, J. Daich and I. Soliman, *et al.*, Vitamin D and autoimmunity, *Scand. J. Rheumatol.*, 2016, **45**(6), 439–447.
14. A. Watad, S. Azrielant and N. L. Bragazzi, *et al.*, Seasonality and autoimmune diseases: The contribution of the four seasons to the mosaic of autoimmunity, *J. Autoimmun.*, 2017, **82**, 13–30.
15. Y. Segal and Y. Shoenfeld, Vaccine-induced autoimmunity: the role of molecular mimicry and immune crossreaction, *Cell Mol. Immunol.*, 2018, **15**(6), 586–594.
16. Y. Deng, X. Huang and H. Wu, *et al.*, Some like it hot: The emerging role of spicy food (capsaicin) in autoimmune diseases, *Autoimmun. Rev.*, 2016, **15**(5), 451–456.
17. G. Aljadeff, E. Longhi and Y. Shoenfeld, Bisphenol A: A notorious player in the mosaic of autoimmunity, *Autoimmunity*, 2018, **51**(8), 370–377.
18. D. Kharrazian, The potential roles of Bisphenol A (BPA) Pathogenesis in Autoimmunity, *Autoimmune Dis.*, 2014, **2014**, 743616.
19. J. A. Rogers, L. Metz and V. W. Yong, Review: Endocrine disrupting chemicals and immune responses: a focus on bisphenol-A and its potential mechanisms, *Mol. Immunol.*, 2013, **53**(4), 421–430.
20. Y. Liu, K. Qu and Y. Hai, *et al.*, Bisphenol A (BPA) binding on full-length architectures of estrogen receptor, *J. Cell. Biochem.*, 2018, **119**(8), 6784–6794.
21. A. Matsushima, Y. Kakuta and T. Teramoto, *et al.*, Structural evidence for endocrine disruptor bisphenol A binding to human nuclear receptor ERR gamma, *J. Biochem.*, 2007, **142**(4), 517–524.
22. H. Nishizawa, M. Morita and M. Sugimoto, *et al.*, Effects of in utero exposure to bisphenol A on mRNA expression of arylhydrocarbon and retinoid receptors in murine embryos, *J. Reprod. Dev.*, 2005, **51**(3), 315–324.
23. C. Park, H. Song and J. Choi, *et al.*, The mixture effects of bisphenol derivatives on estrogen receptor and androgen receptor, *Environ. Pollut.*, 2020, **260**, 114036.
24. P. Gao, L. Wang and N. Yang, *et al.*, Peroxisome proliferator-activated receptor gamma (PPARgamma) activation and metabolism disturbance

induced by bisphenol A and its replacement analog bisphenol S using in vitro macrophages and in vivo mouse models, *Environ. Int.*, 2020, **134**, 105328.
25. H. MacKay and A. Abizaid, A plurality of molecular targets: The receptor ecosystem for bisphenol-A (BPA), *Horm. Behav.*, 2018, **101**, 59–67.
26. F. M. Huang, Y. C. Chang and S. S. Lee, *et al.*, Bisphenol A exhibits cytotoxic or genotoxic potential via oxidative stress-associated mitochondrial apoptotic pathway in murine macrophages, *Food Chem. Toxicol.*, 2018, **122**, 215–224.
27. H. Guo, T. Liu and Y. Uemura, *et al.*, Bisphenol A in combination with TNF-alpha selectively induces Th2 cell-promoting dendritic cells in vitro with an estrogen-like activity, *Cell. Mol. Immunol.*, 2010, 7(3), 227–234.
28. S. L. Liao, M. H. Tsai and S. H. Lai, *et al.*, Prenatal exposure to bisphenol-A is associated with Toll-like receptor-induced cytokine suppression in neonates, *Pediatr. Res.*, 2016, **79**(3), 438–444.
29. M. Yang, W. Qiu and B. Chen, *et al.*, The in vitro immune modulatory effect of bisphenol A on fish macrophages via estrogen receptor α and nuclear factor-κB signaling, *Environ. Sci. Technol.*, 2015, **49**(3), 1888–1895.
30. P. J. Murray and T. A. Wynn, Protective and pathogenic functions of macrophage subsets, *Nat. Rev. Immunol.*, 2011, **11**(11), 723–737.
31. F. M. Huang, Y. C. Chang and S. S. Lee, *et al.*, Expression of pro-inflammatory cytokines and mediators induced by Bisphenol A via ERK-NFκB and JAK1/2-STAT3 pathways in macrophages, *Environ. Toxicol.*, 2019, **34**(4), 486–494.
32. T. Lawrence, The nuclear factor NF-kappaB pathway in inflammation, *Cold Spring Harbor Perspect. Biol.*, 2009, **1**(6), a001651.
33. B. Miraghazadeh and M. C. Cook, Nuclear Factor-kappaB in Autoimmunity: Man and Mouse, *Front. Immunol.*, 2018, **9**, 613.
34. D. Teixeira, C. Marques and D. Pestana, *et al.*, Effects of xenoestrogens in human M1 and M2 macrophage migration, cytokine release, and estrogen-related signaling pathways, *Environ. Toxicol.*, 2016, **31**(11), 1496–1509.
35. A. Wardowska, The epigenetic face of lupus: Focus on antigen-presenting cells, *Int. Immunopharmacol.*, 2020, **81**, 106262.
36. M. H. Biermann, S. Veissi and C. Maueröder, *et al.*, The role of dead cell clearance in the etiology and pathogenesis of systemic lupus erythematosus: dendritic cells as potential targets, *Expert Rev. Clin. Immunol.*, 2014, **10**, 1151–1164.
37. H. Guo, T. Liu and Y. Uemura, *et al.*, Bisphenol A in combination with TNF-α selectively induces Th2 cell-promoting dendritic cells in vitro with an estrogen-like activity, *Cell. Mol. Immunol.*, 2010, 7(3), 227–234.
38. M. Goto, Y. Takano-Ishikawa and H. Ono, *et al.*, Orally administered bisphenol A disturbed antigen specific immunoresponses in the naïve condition, *Biosci., Biotechnol., Biochem.*, 2007, **71**(9), 2136–2143.

39. M. Cernadas, J. Lu and G. Watts, *et al.*, CD1a expression defines an interleukin-12 producing population of human dendritic cells, *Clin. Exp. Immunol.*, 2009, **155**, 523–533.
40. J. Wang, C. She and Z. Li, *et al.*, In vitro impact of bisphenol A on maturation and function of monocyte-derived dendritic cells in patients with primary Sjögren's syndrome, *Immunopharmacol. Immunotoxicol.*, 2020, **42**(1), 28–36.
41. R. Yazdani, M. Sharifi and A. S. Shirvan, *et al.*, Characteristics of innate lymphoid cells (ILCs) and their role in immunological disorders (an update), *Cell. Immunol.*, 2015, **298**(1–2), 66–76.
42. T. Xiong and J. E. Turner, Innate lymphoid cells in autoimmunity and chronic inflammatory diseases, *Semin. Immunopathol.*, 2018, **40**(4), 393–406.
43. Y. Malaisé, S. Ménard and C. Cartier, *et al.*, Consequences of bisphenol a perinatal exposure on immune responses and gut barrier function in mice, *Arch. Toxicol.*, 2018, **92**(1), 347–358.
44. K. H. Lee, A. Kronbichler and D. D. Park, *et al.*, Neutrophil extracellular traps (NETs) in autoimmune diseases: A comprehensive review, *Autoimmun. Rev.*, 2017, **16**(11), 1160–1173.
45. B. Pinegin, N. Vorobjeva and V. Pinegin, Neutrophil extracellular traps and their role in the development of chronic inflammation and autoimmunity, *Autoimmun. Rev.*, 2015, **14**(7), 633–640.
46. J. M. Berthelot, B. Le Goff and A. Neel, *et al.*, NETosis: At the crossroads of rheumatoid arthritis, lupus, and vasculitis, *Jt., Bone, Spine*, 2017, **84**(3), 255–262.
47. H. Meng, S. Yalavarthi and Y. Kanthi, *et al.*, In Vivo Role of Neutrophil Extracellular Traps in Antiphospholipid Antibody-Mediated Venous Thrombosis, *Arthritis Rheumatol.*, 2017, **69**(3), 655–667.
48. S. Tong, S. Yang and T. Li, *et al.*, Role of neutrophil extracellular traps in chronic kidney injury by bisphenol-A, *J. Endocrinol.*, 2019, **241**(2), 125–134.
49. E. Frangou, A. Chrysanthopoulou and A. Mitsios, *et al.*, REDD1/autophagy pathway promotes thromboinflammation and fibrosis in human systemic lupus erythematosus (SLE) through NETs decorated with tissue factor (TF) and interleukin-17A (IL-17A), *Ann. Rheum. Dis.*, 2019, **78**(2), 238–248.
50. L. Doridot, M. Jeljeli and C. Chêne, *et al.*, Implication of oxidative stress in the pathogenesis of systemic sclerosis via inflammation, autoimmunity and fibrosis, *Redox Biol.*, 2019, **25**, 101122.
51. N. R. Gassman, Induction of oxidative stress by bisphenol A and its pleiotropic effects, *Environ. Mol. Mutagen.*, 2017, **58**(2), 60–71.
52. Y. Wang, S. Li and C. Li, Perspectives of New Advances in the Pathogenesis of Vitiligo: From Oxidative Stress to Autoimmunity, *Med. Sci. Monit.*, 2019, **25**, 1017–1023.
53. T. Ozaydin, Y. Oznurlu and E. Sur, *et al.*, Effects of bisphenol A on antioxidant system and lipid profile in rats, *Biotech. Histochem.*, 2018, **93**, 231–238.

54. A. Balistrieri, L. Hobohm and T. Srivastava, *et al.*, Alterations in human neutrophil function by Bisphenol A, *Am. J. Physiol.: Cell Physiol.*, 2018, **315**(5), C636–C642.
55. Y. J. Yang, Bisphenol A exposure is associated with oxidative stress and inflammation in postmenopausal women, *Environ. Res.*, 2009, **109**(6), 797–801.
56. Y. X. Wang, C. Liu and Y. Shen, *et al.*, Urinary levels of bisphenol A, F and S and markers of oxidative stress among healthy adult men: Variability and association analysis, *Environ. Int.*, 2019, **123**, 301–309.
57. H. Vaseghi, M. Houshmand and Z. Jadali, Increased levels of mitochondrial DNA copy number in patients with vitiligo, *Clin. Exp. Dermatol.*, 2017, **42**(7), 749–754.
58. H. Xie, F. Zhou and L. Liu, *et al.*, Vitiligo: How do oxidative stress-induced autoantigens trigger autoimmunity?, *J. Dermatol. Sci.*, 2016, **81**(1), 3–9.
59. P. Miossec and J. K. Kolls, Targeting IL-17 and TH17 cells in chronic inflammation, *Nat. Rev. Drug Discovery*, 2012, **11**(10), 763–776.
60. K. D. Moudgil and D. Choubey, Cytokines in autoimmunity: role in induction, regulation, and treatment, *J. Interferon Cytokine Res.*, 2011, **3**(10), 695–703.
61. H. Yan, M. Takamoto and K. Sugane, Exposure to Bisphenol A prenatally or in adulthood promotes T(H)2 cytokine production associated with reduction of CD4CD25 regulatory T cells, *Environ. Health Perspect.*, 2008, **116**, 514–519.
62. M. Goto, Y. Takano-Ishikawa and H. Ono, *et al.*, Orally administered Bisphenol A disturbed antigen specific immunoresponses in the naive condition, *Biosci., Biotechnol., Biochem.*, 2007, **71**(9), 2136–2143.
63. J. S. Ellis, S. H. Hong and H. Zaghouani, *et al.*, Reduced effectiveness of CD4+Foxp3+ regulatory T cells in CD28-deficient NOD.H-2h4 mice leads to increased severity of spontaneus autoimmune thyroiditis, *J. Immunol.*, 2013, **191**(10), 4940–4949.
64. S. Yoshino, K. Yamaki and X. Li, *et al.*, Prenatal exposure to bisphenol A up-regulates immune responses, including T helper 1 and T helper 2 responses, in mice, *Immunology*, 2004, **112**(3), 489–495.
65. Y. Liu, C. Mei and H. Liu, *et al.*, Modulation of cytokine expression in human macrophages by endocrine-disrupting chemical bisphenol A, *Biochem. Biophys. Res. Commun.*, 2014, **451**, 592–598.
66. J. M. Gostner, E. Raggl and K. Becker, *et al.*, Bisphenol A suppresses Th1-type immune response in human peripheral blood mononuclear cells in vitro, *Immunol. Lett.*, 2015, **168**(2), 285–292.
67. H. Yan, M. Takamoto and K. Sugane, Exposure to Bisphenol A prenatally or in adulthood promotes T(H)2 cytokine production associated with reduction of CD4CD25 regulatory T cells, *Environ. Health Perspect.*, 2008, **116**(4), 514–519.
68. H. Guo, T. Liu and Y. Uemura, *et al.*, Bisphenol A in combination with TNF-alpha selectively induces Th2 cell-promoting dendritic cells

in vitro with an estrogen-like activity, *Cell. Mol. Immunol.*, 2010, 7(3), 227–234.
69. S. Luo, Y. Li and Y. Li, *et al.*, Gestational and lactational exposure to low-dose bisphenol A increases Th17 cells in mice offspring, *Environ. Toxicol. Pharmacol.*, 2016, **47**, 149–158.
70. G. Wang, Y. Li and Y. Li, *et al.*, Maternal vitamin D supplementation inhibits bisphenol A-induced proliferation of Th17 cells in adult offspring, *Food Chem. Toxicol.*, 2020, **144**, 111604.
71. M. H. Lee, W. Chung and B. Y. Kang, *et al.*, Enhanced interleukin-4 production in CD4+ T-cells and elevated immunoglobulin E levels in antigen-primed mice by Bisphenol A and nonylphenol, endocrine disruptors: involvement of nuclear factor-AT and CA2+, *Immunology*, 2003, **109**(1), 76–86.
72. I. J. Wang, C. Y. Chen and C. G. Bornehag, Bisphenol A exposure may increase the risk of development of atopic disorders in children, *Int. J. Hyg. Environ. Health*, 2016, **219**(3), 311–316.
73. M. Maurer, S. Altrichter and O. Schmetzer, *et al.*, Immunoglobulin E-Mediated Autoimmunity, *Front. Immunol.*, 2018, **9**, 689.
74. M. A. Sanjuan, D. Sagar and R. Kolbeck, Role of IgE in autoimmunity, *J. Allergy Clin. Immunol.*, 2016, **137**(6), 1651–1661.
75. M. García-Carrasco, S. Macias-Díaz and C. Mendoza-Pinto, *et al.*, The role of IgE in systemic lupus erythematosus, *Autoimmun. Rev.*, 2021, **20**(1), 102704.
76. A. M. Atta, M. B. Santiago and F. G. Guerra, *et al.*, Autoimmune response of IgE antibodies to cellular self-antigens in systemic Lupus Erythematosus, *Int. Arch. Allergy Immunol.*, 2010, **152**(4), 401–406.
77. B. Dema, N. Charles and C. Pellefigues, *et al.*, Immunoglobulin E plays an immunoregulatory role in lupus, *J. Exp. Med.*, 2014, **211**(11), 2159–2168.
78. I. Sekigawa, K. Kaneda and H. Kaneko, *et al.*, Detection of serum IgE class anti-SSA antibodies in mothers with foetal loss, *Rheumatol. Int.*, 2008, **28**, 623–626.
79. R. A. Ferreira, V. P. Ferriani and M. C. Sopelete, *et al.*, Immunoglobulin E-rheumatoid factor in juvenile rheumatoid arthritis, *Rev. Hosp. Clin., Fac. Med. Univ. Sao Paulo*, 2002, **57**(5), 209–216.
80. H. Yurino, S. Ishikawa and T. Sato, *et al.*, Endocrine disruptors (environmental estrogens) enhance autoantibody production by B1 cells, *Toxicol. Sci*, 2004, **81**(1), 139–147.
81. S. Morimoto, S. Nakano and T. Watanabe, *et al.*, Expression of B-cell activating factor of the tumour necrosis factor family (BAFF) in T cells in active systemic lupus erythematosus: the role of BAFF in T cell-dependent B cell pathogenic autoantibody production, *Rheumatology*, 2007, **46**(7), 1083–1086.
82. U. Steffen, C. A. Koeleman and M. V. Sokolova, *et al.*, IgA subclasses have different effector functions associated with distinct glycosylation profiles, *Nat. Commun.*, 2020, **11**(1), 120.

83. A. Jiménez, R. Alvarez-Doforno and M. C. García Rodríguez, *et al.*, Autoantibodies in patients with IgA and IgG2 deficiencies, *Apmis*, 1991, **99**(4), 327–332.
84. T. P. Primm and H. F. Gilbert, Hormone binding by protein disulphide isomerase, a high capacity hormone reservoir or the endoplasmic reticulum, *J. Biol. Chem.*, 2001, **276**(1), 281–286.
85. S. Hashimoto, K. Shiomoto and K. Okada, *et al.*, The binding site of bisphenol A to protein disulphide isomerase, *J. Biochem.*, 2012, **151**(1), 35–45.
86. T. Hiroi, K. Okada and S. Imaoka, *et al.*, Bisphenol A binds to protein disulfide isomerase and inhibits its enzymatic and hormone-binding activities, *Endocrinology*, 2006, **147**(6), 2773–2780.
87. D. Kharrazian, M. Herbert and A. Vojdani, The Associations between Immunological Reactivity to the Haptenation of Unconjugated Bisphenol A to Albumin and Protein Disulfide Isomerase with Alpha-Synuclein Antibodies, *Toxics*, 2019, 7(2), 26.
88. D. Kharrazian and A. Vojdani, Correlation between antibodies to bisphenol A, its target enzyme protein disulfide isomerase and antibodies to neuron-specific antigens, *J. Appl. Toxicol.*, 2017, **37**(4), 479–484.
89. A. Vojdani, D. Kharrazian and P. S. Mukherjee, Elevated levels of antibodies against xenobiotics in a subgroup of healthy subjects, *J. Appl. Toxicol.*, 2015, **35**(4), 383–397.
90. D. Kharrazian, M. Martha Herbert and Aristo, Vojdani A. The Associations between Immunological Reactivity to the Haptenation of Unconjugated Bisphenol A to Albumin and Protein Disulfide Isomerase with Alpha-Synuclein Antibodies, *Toxics*, 2019, 7(2), 26.
91. N. Molodecky and G. Kaplan, Environmental risk factors for inflammatory bowel disease, *Gastroenterol. Hepatol.*, 2010, **6**, 339–346.
92. E. Legaki and M. Gazouli, Influence of environmental factors in the development of inflammatory bowel disease, *World J. Gastrointest. Pharmacol. Ther.*, 2016, 7, 112.
93. J. Xu, G. Huang and T. Nagy, *et al.*, Guo TL. Bisphenol A Alteration Of Type 1 Diabetes In Non-obese Diabetic (NOD) Female Mice Is Dependent On Window Of Exposure, *Arch. Toxicol.*, 2019, **93**(4), 1083–1093.
94. A. Javurek, W. Spollen and S. Johnson, *et al.*, Effects of exposure to bisphenol A and ethinyl estradiol on the gut microbiota of parents and their offspring in a rodent model, *Gut Microbes*, 2016, 7, 471–485.
95. K. Moryiama, T. Tagami and T. Akamizu, *et al.*, Thyroid hormone action is disrupted by Bisphenol A as an antagonist, *J. Clin. Endocrinol. Metab.*, 2002, **87**(11), 5185–5190.
96. J. M. Davies, Molecular mimicry: can epitope mimicry induce autoimmune disease?, *Immunol. Cell Biol.*, 1997, **75**(2), 113–126.
97. V. V. Borba, G. Zandman-Goddard and Y. Shoenfeld, Prolactin and Autoimmunity, *Front. Immunol.*, 2018, **9**, 73.

98. V. V. Borba, G. Zandman-Goddard and Y. Shoenfeld, Prolactin and autoimmunity: The hormone as an inflammatory cytokine, *Best Pract. Res., Clin. Endocrinol. Metab.*, 2019, **33**(6), 101324.
99. R. Steinmetz, N. G. Brown and D. L. Allen, *et al.*, The environmental estrogen bisphenol A stimulated prolactin release in vitro and in vivo, *Endocrinology*, 1997, **138**, 1780–1786.
100. L. Hao, J. Zhang and Y. Zhang, *et al.*, Effect of bisphenol a on occurrence and progression of prolactinoma and its underlying mechanisms, *Am. J. Transl. Res.*, 2016, **8**(10), 4195–4204.
101. R. Steinmetz, N. G. Brown and D. L. Allen, *et al.*, The environmental estrogen bisphenol A stimulates prolactin release in vitro and in vivo, *Endocrinology*, 1997, **138**(5), 1780–1786.
102. E. Peeva and M. Zouali, Spotlight on the role of hormonal factors in the emergence of autoreactive B-lymphocytes, *Immunol. Lett.*, 2005, **101**, 123–143.
103. A. Watad, H. Amital and G. Aljadeff, *et al.*, Prolactin, another important player in the mosaic of autoimmunity, *Isr. Med. Assoc. J.*, 2016, **18**, 542–543.
104. E. Peeva, D. Michael and J. Cleary, *et al.*, Prolactin modulates the naive B cell repertoire, *J. Clin. Invest.*, 2003, **111**(2), 275–283.
105. L. J. Jara, G. Medina and M. A. Saavedra, *et al.*, Prolactin and autoimmunity, *Clin. Rev. Allergy Immunol.*, 2011, **40**, 50–59.
106. D. G. Doherty, Immunity, tolerance and autoimmunity in the liver: a comprehensive review, *J. Autoimmun.*, 2016, **66**, 60–75.
107. M. R. Namazi, Cytochrome-P450 enzymes and autoimmunity: expansion of the relationship and introduction of free radicals as the link, *J. Autoimmune Dis.*, 2009, **6**, 4–6.
108. N. R. Gassman, Induction of oxidative stress by bisphenol A and its pleiotropic effects, *Environ. Mol. Mutagen.*, 2017, **58**, 60–71.
109. J. Xu, G. Huang and T. Guo, Developmental Bisphenol A Exposure Modulates Immune-Related Diseases, *Toxics*, 2016, **4**(4), 23.
110. M. Kundakovic, K. Gudsnuk and B. Franks, *et al.*, Sex-specific epigenetic disruption and behavioral changes following low-dose in utero bisphenol A exposure, *Proc. Natl. Acad. Sci. U. S. A.*, 2013, **110**(24), 9956–9961.
111. A. Veiga-Lopez, *et al.*, Developmental programming: gestational bisphenol-A treatment alters trajectory of fetal ovarian gene expression, *Endocrinology*, 2013, **154**, 1873–1874.
112. S. Singh and S. S. Li, Epigenetic Effects of Environmental Chemicals Bisphenol A and Phthalates, *Int. J. Mol. Sci.*, 2012, **13**(8), 10143–10153.
113. D. Khan and S. A. Ahmed, Epigenetic Regulation of Non-Lymphoid Cells by Bisphenol A, a Model Endocrine Disrupter: Potential Implications for Immunoregulation, *Front. Endocrinol.*, 2015, **6**, 91.
114. H. T. Alhomaidan, N. Rasheed and S. Almatrafi, *et al.*, Bisphenol A modified DNA: A possible immunogenic stimulus for anti-DNA autoantibodies in systemic lupus erythematosus, *Autoimmunity*, 2019, **52**(7–8), 272–280.

115. Y. Dong, Z. Zhang and H. Liu, *et al.*, Exacerbating lupus nephritis following BPA exposure is associated with abnormal autophagy in MRL/lpr mice, *Am. J. Transl. Res.*, 2020, **12**(2), 649–659.
116. M. Sugiura-Ogasawara, Y. Ozaki and S. Sonta, *et al.*, Exposure to bisphenol A is associated with recurrent miscarriage, *Hum. Reprod.*, 2005, **20**(8), 2325–2329.
117. J. Wang, C. She and Z. Li, *et al.*, In vitro impact of bisphenol A on maturation and function of monocyte-derived dendritic cells in patients with primary Sjögren's syndrome, *Immunopharmacol. Immunotoxicol.*, 2020, **42**(1), 28–36.
118. E. Palioura and E. Diamanti-Kandarakis, Polycystic ovary syndrome (PCOS) and endocrine disrupting chemicals (EDCs), *Rev. Endocr. Metab. Disord.*, 2015, **16**(4), 365–371.
119. E. C. Bonefeld-Jørgensen, M. Long and M. V. Hofmeister, *et al.*, Endocrine-disrupting potential of bisphenol A, bisphenol A dimethacrylate, 4-n-nonylphenol, and 4-n-octylphenol in vitro: new data and a brief review, *Environ. Health Perspect.*, 2007, **115**(Suppl 1), 69–76.
120. K. S. Kechagias, A. Semertzidou and A. Athanasiou, *et al.*, Bisphenol-A and polycystic ovary syndrome: a review of the literature, *Rev. Environ. Health*, 2020, **35**(4), 323–331.
121. C. Pivonello, G. Muscogiuri and A. Nardone, *et al.*, Bisphenol A: an emerging threat to female fertility, *Reprod. Biol. Endocrinol.*, 2020, **18**(1), 22.
122. M. O. Fernandez, N. S. Bourguignon and P. Arocena, *et al.*, Neonatal exposure to bisphenol A alters the hypothalamic-pituitary-thyroid axis in female rats, *Toxicol. Lett.*, 2018, **285**, 81–86.
123. R. Chianese, A. Viggiano and K. Urbanek, *et al.*, Chronic exposure to low dose of bisphenol A impacts on the first round of spermatogenesis via SIRT1 modulation, *Sci. Rep.*, 2018, **8**(1), 2961.
124. R. T. Zoeller, R. Bansal and C. Parris, Bisphenol-A, an environmental contaminant that acts as a thyroid hormone receptor antagonist in vitro, increases serum thyroxine, and alters RC3/neurogranin expression in the developing rat brain, *Endocrinology*, 2005, **146**(2), 607–612.
125. M. J. Kim and Y. J. Park, Bisphenols and Thyroid Hormone, *Endocrinol. Metab.*, 2019, **34**, 340–348.
126. J. D. Meeker and K. K. Ferguson, Relationship between urinary phtalate and bisphenol A concentrations and serum thyroid measures in U.S. adults and adolescents from the National Health and Nutrition Examination Survey (NHANES) 2007-2008, *Environ. Health Perspect.*, 2011, **119**, 1396–1402.
127. C. Sriphrapradang, L. O. Chailurkit and W. Aekplakorn, *et al.*, Association between bisphenol A and abnormal free thyroxine level in men, *Endocrine*, 2013, **44**(2), 441–447.
128. T. Wang, J. Lu and M. Xu, *et al.*, Urinary bisphenol A concentration and thyroid function in Chinese adults, *Epidemiol.*, 2013, **24**, 295–302.

129. C. Park, W. Choi and M. Hwang, *et al.*, Associations between urinary phtalate metabolites and bisphenol A levels, and serum thyroid hormones among the Korean adult population: Korean National Environmetal Health Survey (KoNEHS) 2012-2014, *Sci. Total Environ*, 2017, **584–585**, 950–957.
130. Z. Zhou, J. Zhang and F. Jiang, *et al.*, Higher urinary bisphenol A concentration and excessive iodine intake are associated with nodular goiter and papillary thyroid carcinoma, *Biosci. Rep.*, 2017, **37**, 1–10.
131. L. Li, Y. Ying and C. Zhang, *et al.*, Bisphenol A exposure and risk of thyroid nodules in Chinese women: a case-control study, *Environ. Int.*, 2019, **126**, 321–328.
132. X. Wang, N. Tang and S. F. Nakayama, *et al.*, Maternal urinary bisphenol A concentration and thyroid hormone levels of Chinese mothers and newborns by maternal body mass index, *Environ. Sci. Pollut. Res. Int.*, 2020, **27**(10), 10939–10949.
133. L. O. Chailurkit, W. Aekplakorn and B. Ongphiphadhanakul, The Association of Serum Bisphenol A with Thyroid Autoimmunity, *Int. J. Environ. Res. Public Health*, 2016, **13**(11), 1153.
134. N. Maserejian, Possible association between dental sealants and urinary bisphenol A levels in children warrants additional biomonitoring and safety research, *J. Evid. Based Dent. Pract.*, 2014, **4**(4), 200–202.
135. P. Vabre, N. Gatimel and J. Moreau, *et al.*, Environmental pollutants, a possible etiology for premature ovarian insufficiency: a narrative review of animal and human data, *Environ. Health*, 2017, **16**(37), 1–18.
136. S. Patel, C. Zhou and S. Rattan, *et al.*, Effects of endocrine-disrupting chemicals on the ovary, *Biol. Reprod.*, 2015, **93**(20), 1–9.
137. S. G. Howard, Exposure to environmental chemicals and type 1 diabetes: an update, *J. Epidemiol. Community Health*, 2019, **73**(6), 483–488.
138. J. Bodin, L. C. Stene and U. C. Nygaard, Can exposure to environmental chemicals increase the risk of diabetes type 1 development?, *BioMed Res. Int.*, 2015, **2015**, 208947.
139. J. Bodin, A. Kocbach Bølling and A. Wendt, *et al.*, Exposure to bisphenol A, but not phthalates, increases spontaneous diabetes type 1 development in NOD mice, *Toxicol. Rep.*, 2015, **2**, 99–110.
140. J. Xu, G. Huang and T. Nagy, *et al.*, Bisphenol A alteration of type 1 diabetes in non-obese diabetic (NOD) female mice is dependent on window of exposure, *Arch. Toxicol.*, 2019, **93**(4), 1083–1093.
141. C. Ahn, H. S. Kang and J. H. Lee, *et al.*, Bisphenol A and octylphenol exacerbate type 1 diabetes mellitus by disrupting calcium homeostasis in mouse pancreas, *Toxicol. Lett.*, 2018, **295**, 162–172.
142. T. Ince, A. Balcı and S. S. Yalçın, *et al.*, Urinary bisphenol-A levels in children with type 1 diabetes mellitus, *J. Pediatr. Endocrinol. Metab.*, 2018, **31**(8), 829–836.
143. S. Yan, W. Song and Y. Chen, *et al.*, Low-dose bisphenol A and estrogen increase ventricular arrhythmias following ischemia-reperfusion in female rat hearts, *Food Chem. Toxicol.*, 2013, **56**, 75–80.

144. D. Melzer, P. Gates and N. J. Osborne, *et al.*, Urinary bisphenol a concentration and angiography-defined coronary artery stenosis, *PLoS One*, 2012, 7(8), e43378.
145. X. Gao and H. S. Wang, Impact of bisphenol A on the cardiovascular system – epidemiological and experimental evidence and molecular mechanisms, *Int. J. Environ. Res. Public Health*, 2014, **11**, 8399–8413.
146. Z. Lazurova and I. Tkac, Bisfenol A a jeho vplyv na kardiovaskulárny system, *Int. Med.*, 2015, **15**(2), 62–64.
147. K. A. Bruno, J. E. Mathews and A. L. Yang, *et al.*, BPA Alters Estrogen Receptor Expression in the Heart After Viral Infection Activating Cardiac Mast Cells and T Cells Leading to Perimyocarditis and Fibrosis, *Front. Endocrinol.*, 2019, **10**(598), 1–16.
148. S. M. Belcher, Y. Chen and S. Yan, *et al.*, Rapid estrogen receptor-mediated mechanisms determine the sexually dimorphic sensitivity of ventricular myocytes to 17beta-estradiol and the environmental endocrine disruptor bisphenol A, *Endocrinology*, 2012, **153**(2), 712–720.
149. A. Koenig, I. Buskiewicz and S. A. Huber, Age-associated changes in estrogen receptor ratios correlate with increased female susceptibility to coxackievirus B3-induced myocarditis, *Front. Immunol.*, 2017, **8**, 1585.
150. K. Stafford, H. Greyner and F. Molina, *et al.*, Bisphenol A (BPA) exacerbates acute myocarditis in female BALB/c by activating mast cells and the inflammasome (HUM1P.302), *J. Immunol.*, 2014, **192**(Suppl. 1), 52-2.
151. K. P. Lai, Y. T. Chung and R. Li, *et al.*, Bisphenol A alters gut microbiome: Comparative metagenomics analysis, *Environ. Pollut.*, 2016, **218**, 923–930.
152. J. A. A. DeLuca, K. F. Allred and R. Menon, *et al.*, Bisphenol-A alters microbiota metabolites derived from aromatic amino acids and worsens disease activity during colitis, *Exp. Biol. Med.*, 2018, **243**(10), 864–875.
153. D. Owczarek, T. Rodacki and R. Domagała-Rodacka, *et al.*, Diet and nutritional factors in inflammatory bowel diseases, *World J. Gastroenterol.*, 2016, **22**(3), 895–905.
154. P. S. De Silva, X. Yang and J. R. Korzenik, *et al.*, Association of urinary phenolic compounds, inflammatory bowel disease and chronic diarrheal symptoms: Evidence from the National Health and Nutrition Examination Survey, *Environ. Pollut.*, 2017, **229**, 621–626.
155. J. A. Rogers, M. K. Mishra and J. Hahn, *et al.*, Gestational bisphenol-A exposure lowers the threshold for autoimmunity in a model of multiple sclerosis, *Proc. Natl. Acad. Sci. U. S. A.*, 2017, **114**(19), 4999–5004.
156. D. N. Krementsov, A. Katchy and K. Laure, *et al.*, Studies in Experimental Autoimmune Encephalomyelitis Do Not Support Developmental Bisphenol A Exposure as an Environmental Factor in Increasing Multiple Sclerosis Risk, *Toxicol. Sci*, 2013, **135**(1), 91–102.
157. C. Brinkmeyer-Langford, A. Rodrigues and K. J. Kochan, *et al.*, Consequences of perinatal bisphenol A exposure in a mouse model of multiple sclerosis, *Autoimmunity*, 2014, **47**(1), 55–56.

158. D. Kharrazian and A. Vojdani, Correlation between antibodies to bisphenol A, its target enzyme protein disulfide isomerase and antibodies to neuron-specific antigens, *J. Appl. Toxicol.*, 2017, **37**(4), 479–484.
159. F. Garretti, D. Agalliu and A. Lindestam, *et al.*, Autoimmunity in Pakinson's disease: The role of alpha-synuclein-specific T cells, *Front. Immunol.*, 2019, **10**, 303.
160. D. Kharrazian, M. Herbert and A. Vojdani, The associations between immunological reactivity to the haptenation of unconjugated bisphenol A to albumin and protein disulfide isomerase with alpha-synuclein antibodies, *Toxics*, 2019, 7(2), 26.
161. A. Watad, K. Sharif and Y. Shoenfeld, The ASIA syndrome: basic concepts, Mediterranean, *J. Rheumatol.*, 2017, **28**(2), 64–69.
162. A. Watad, P. David and S. Brown, *et al.*, Autoimmune/Inflammatory Syndrome Induced by Adjuvants and Thyroid Autoimmunity, *Front. Endocrinol.*, 2017, 7, 150.

CHAPTER 6

A Novel Model for the Estrogenic Action of BPA in Developing Brain Following Maternal Ingestion

CHRISTINE K. WAGNER*[a] AND ALLYSSA FAHRENKOPF[a,b]

[a] Department of Psychology and Center for Neuroscience Research, University at Albany, Albany, NY 12222, USA; [b] Psychogenics Inc., Paramus, NJ 07652, USA
*Email: cwagner@albany.edu

6.1 Bisphenol A: A Prevalent Environmental Endocrine Disruptor

Bisphenol A (BPA) has been commonly used in the manufacture of polycarbonate plastics, epoxy resins, and flame retardants, and can leach out of many household products such as water bottles, plastic water pipes, thermal receipt paper, and food cans, making it an ever-present chemical contaminant in daily life. Indeed, BPA and its conjugates are detected in the urine of >90% of adults in the United States,[1–3] and unconjugated BPA is routinely detected in blood (reviewed by Vandenberg *et al.*[3]). Even so-called "safe alternatives" to BPA, namely the analogs bisphenol F (BPF) and bisphenol S (BPS), have been detected in the urine of 57.1% and 88.4% of adults in the United States, respectively.[4]

The literature well supports the characterization of BPA as an environmental endocrine disruptor. Although there are numerous molecular pathways by

Issues in Toxicology No. 43
Bisphenol A: A Multi-modal Endocrine Disruptor
Edited by Natalie R. Gassman

Published by the Royal Society of Chemistry, www.rsc.org

which BPA may exert its effects on a variety of tissues, its ability to mimic and/or disrupt estrogen action has been well established. BPA exerts its effects, at least in part, through the activation of the nuclear estrogen receptor α (ERα) and ERβ.[5–10] BPA selectively binds to both receptors, but has a relatively low affinity compared to estradiol.[6,7] Despite this, very low doses of BPA have been associated with adverse effects in humans and animal models, reviewed recently by Patisaul[11] and vom Saal and Vandenberg.[12]

BPA activates molecular mechanisms within ERα that are distinct from the actions of estradiol. This likely results from the different interactions of BPA with the ligand-binding domain of the receptor, which induce a different conformational change than activation by estradiol.[6] In this regard, BPA may activate a unique set of ERα-responsive genes. In addition, the ratio of ERα to ERβ expression within a tissue can influence which ER-responsive genes are induced by BPA exposure.[6] BPA may also exert estrogenic effects in the absence of estradiol, but antagonistic effects at the ER when estradiol is present.[8] These results suggest that bisphenols may be more than just simply estrogen mimics, but rather indicate that their effects may be anatomically and temporally specific, making elucidation of their mechanism of action, particularly during development, even more critical.

6.2 BPA Exposure During Development: The Vulnerable Brain

During development, exposure to BPA is of particular concern, as infants and pregnant and lactating women may ingest BPA. BPA has been detected in human embryos as early as 8 weeks post-fertilization.[13] BPA is measured in maternal urine and serum during pregnancy and is detectable in fetal serum, amniotic fluid, and placental tissue.[14–17] Indeed, BPA was transferred from the maternal compartment to the fetal compartment of the placenta *in situ*, suggesting that BPA is indeed passed from mother to fetus.[18] Interestingly, BPA concentrations in fetal plasma were higher in males than females,[15] creating the potential for differential effects based on gender.

Exposure to BPA during neonatal life is of great concern. BPA has been detected in human breast milk[19,20] and colostrum,[21] suggesting it that could be passed from mother to nursing infant. Additionally, infants and children are likely to ingest BPA directly from food containers and plastic baby bottles. Indeed, not only is BPA detectable in urinary samples from children, but concentrations are higher than those measured in either adults or adolescents.[1] Developmental exposure to BPA can adversely affect many aspects of maturing physiology, but the developing brain may be particularly vulnerable to its harmful effects.

A rapidly growing literature demonstrates that developmental exposure to BPA in animal models can disrupt fundamental mechanisms of brain development (for recent reviews, see ref. 11, 22, 23). The neurodevelopmental effects of BPA exposure are wide-ranging, but include alterations in cortical organization and connectivity,[24–26] inhibition of neurogenesis in perinatal dentate gyrus,[27] and dendritic and/or synaptic development in hypothalamus,[28] cerebellum,[29] and hippocampus.[30] Recent evidence suggests that prenatal BPA exposure produced permanent changes in gene expression profiles in the brain, but did so in a region- and sex-dependent manner.[31] By altering brain development in this way, BPA can impair subsequent adult behavior. For example, exposure during development has been shown to alter sexual behaviors, learning and memory, anxiety, maternal behavior, and social behaviors,[32,33] including aggressive behavior in animal models[34] and humans.[35,36]

The sexual differentiation of the developing brain is highly sensitive to the actions of steroid hormones, and indeed, perinatal exposure to BPA in rodents has been shown to alter sex differences in brain and behavior. Exposure to BPA during early neonatal development disrupted sex differences in dopaminergic neurons and in the kisspeptin cell population in the anterolateral periventricular nucleus,[37,38] a brain region critical for the regulation of gonadotropin-releasing hormone neurons and reproductive function. BPA exposure altered calbindin cell number in the medial preoptic nucleus,[39] a well-known biomarker for steroid-dependent sexual differentiation of this region, regulating sex-dependent behaviors such as reproductive behaviors and aggression. There are also often differential effects of BPA on behavior depending on the sex of the animal[40] or on the gender of the child.[35]

6.3 A Sensitive "Bioassay" for Estrogenic Activity in the Developing Brain

Despite an abundance of convincing evidence to demonstrate that exposure to BPA during development alters brain function and behavior, the mechanisms by which BPA exerts its effects in the developing brain are not well understood. A bioassay for ERα activity in developing brain would be beneficial in this regard. We have introduced such a bioassay in the expression of the nuclear steroid receptor, progesterone receptor (PR), in the perinatal medial preoptic nucleus (MPN). We posit that the expression of PR in the developing MPN of rats provides an innovative and sensitive model system to specifically test estrogenic, ERα-dependent actions of BPA in the fetal brain following maternal ingestion.

Within the developing brain, there is a dramatic sex difference in the expression of PR in the rodent MPN during perinatal life (see Figure 6.1), in which males express high levels of PR beginning around embryonic day 18 (E18), while in females PR expression is virtually absent until the second postnatal week,[41] producing one of the largest and developmentally earliest sex differences observed in the mammalian brain to date. *In vitro* models demonstrate that the expression of the PR gene is driven by estradiol activation of ERα, which, in turn, interacts with a consensus estrogen response element (ERE) within the promoter region of the PR gene to induce transcription.[42,43] Ample evidence indicates that PR expression in the developing MPN is regulated similarly; in effect, estradiol, aromatized from testosterone, acts at ERα to induce PR expression. Indeed, prior to E19, neither males

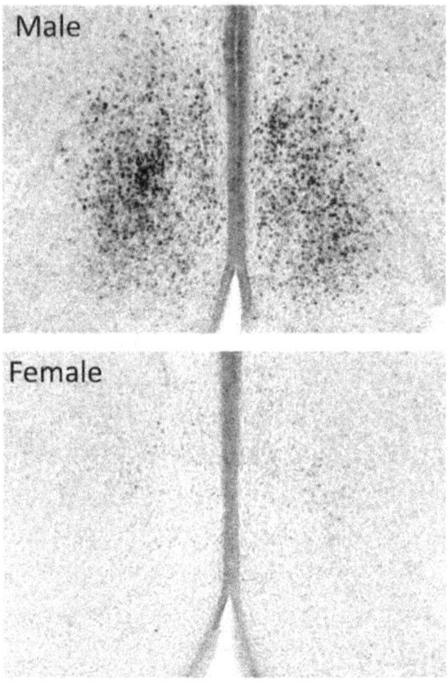

Figure 6.1 Progesterone receptor immunoreactivity in the medial preoptic nucleus of male and female rats on the day of birth.
Reproduced from ref. 41 with permission from Oxford University Press, Copyright 1998.

nor females express detectable levels of PR immunoreactivity in the MPN.[41] However, by E20, following the surge in testosterone secretion from the fetal testes, males begin to express higher levels of PR, whereas females fail to express PR in the absence of ovarian hormone secretion prenatally. Castration of males on the day of birth abolished PR expression in the MPN.[44] PR expression was reduced in males following administration of the aromatase inhibitor androstenedione (ATD) prenatally,[45] verifying that it is indeed estradiol that is the active steroid hormone in inducing PR.

Conversely, PR expression can be readily induced in the perinatal female MPN by exogenous estrogens. For example, prenatal administration of testosterone or the synthetic estrogen, diethylstilbestrol (DES), but not dihydrotestosterone, significantly increased PR expression in the fetal female MPN.[45] Estradiol benzoate treatment in neonatal females increased PR expression, but this effect was completely abolished by the concurrent administration of the ER antagonists tamoxifen or ICI 182,780.[46] PR expression in the perinatal MPN is specifically dependent on ERα. Neonatal administration of the ERα agonist, propyl pyrazole triol (PPT), but not the ERβ agonist, diarylpropionitrile (DPN), increased levels of PR immunoreactivity in the MPN of postnatal females.[47] The sex difference in PR expression in the MPN is completely abolished in ERα-knockout mice, with PR expression essentially absent in males and females lacking a functional ERα gene.[48]

Consequently, we suggest that PR expression in the perinatal MPN of male and female rats provides an innovative and highly sensitive model system to examine estrogenic, ERα-dependent actions of BPA on the developing brain following maternal ingestion.

6.4 A Novel Model System for ERα-dependent Actions of BPA in Developing Brain

6.4.1 Maternal Ingestion of BPA During Gestation Induces PR in Female MPN

In a test of this model system, the induction of PR immunoreactivity in the MPN of fetal females was assessed following maternal ingestion of BPA during pregnancy.[49] Pregnant females received 10 $\mu g\,kg^{-1}$ (low dose) or 50 $\mu g\,kg^{-1}$ (high dose) BPA on a vanilla wafer or received a plain vanilla wafer each day on gestational days 15 through 22. These doses (10 and 50 $\mu g\,kg^{-1}$ per day) are based on the United States Enivronmental Protection Agency maximum acceptable dose of 50 $\mu g\,kg^{-1}$ per day and the European Union's temporary tolerable daily intake of 10 $\mu g\,kg^{-1}$ per day). Pups were delivered normally on gestational day 22, and sections through the MPN of female neonates were collected on the day of birth. Immunocytochemistry for PR was performed using a primary antiserum (Dako Corp.) that recognizes both PR_A and PR_B isoforms. Maternal ingestion of the low dose of BPA significantly increased levels of PR immunoreactivity in the neonatal female MPN compared to controls (Figure 6.2). Interestingly, the higher dose failed to induce PR expression over control levels. These findings are consistent with the idea that BPA ingested by the pregnant female exerts estrogenic effects on ERα in the fetal brain and demonstrates the sensitivity of PR expression in MPN as a bioassay for BPA. Our results are also consistent with numerous reports of non-monotonic effects of endocrine disruptors,[3,50–52] and support the notion that lower doses may exert more powerful effects on the fetal brain than seemingly more dangerous higher exposure levels.

6.4.2 BPA Exerts Estrogenic Action on ERα to Induce PR

To determine if the actions of BPA on PR expression in the developing MPN were dependent on ERα activity, a second study administered BPA directly to neonatal females in the presence or absence of an ER antagonist.[49] In this study, BPA was administered directly to female neonates rather than to fetuses *via* maternal ingestion, as the ER antagonist could potentially compromise the pregnancy and possibly interfere with normal parturition. Therefore, neonatal females received daily injections of BPA (5 $\mu g\,kg^{-1}$, subcutaneous) or the vehicle alone on postnatal days 2–4. In addition, females received injections of the ER antagonist, ICI 182,780 (ICI; 1.5 $mg\,kg^{-1}$) or the dimethyl sulfoxide vehicle alone from postnatal days 1 to 4. While there is some controversy about whether or not ICI can pass through the blood–brain barrier, we have demonstrated that ICI can pass through the immature blood–brain barrier during the first postnatal week and act as a

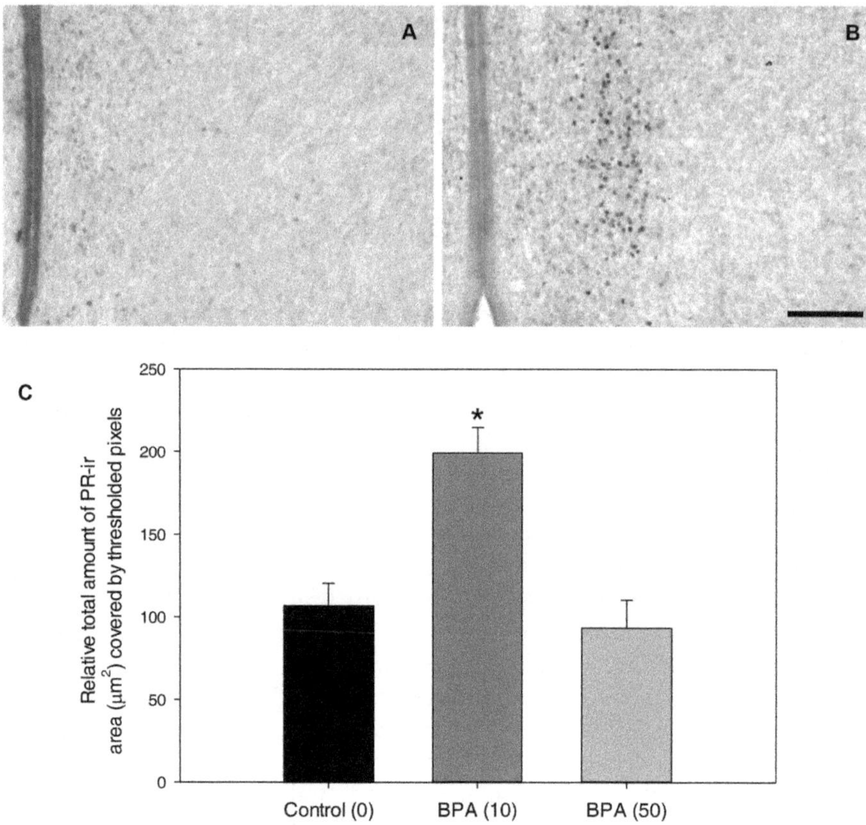

Figure 6.2 (A, B) Progesterone receptor immunoreactivity (PRir) in the medial preoptic nucleus (MPN) of female rats on the day of birth whose mothers ingested (A) control vehicle or (B) bisphenol A (BPA; 10 $\mu g\,kg^{-1}$) from gestational day 15 to 22. Scale bar = 100 μm. (C) The relative total amount of PRir in the MPN of day of birth females whose mothers ingested control vehicle (0), 10 $\mu g\,kg^{-1}$ BPA (10), or 50 $\mu g\,kg^{-1}$ BPA (50) from gestational day 15 to 22. *: significantly different from control and BPA (50) ($p < 0.001$).
Reproduced from ref. 49 with permission from Elsevier, Copyright 2020.

potent antagonist to ERα function in the MPN during the first postnatal week.[46] As in the previous experiment, administration of BPA significantly increased PR expression in the neonatal female MPN on postnatal day 5 compared to controls. However, concurrent administration of ICI abolished the induction of PR by BPA, whereas ICI had no effect in the absence of BPA administration (Figure 6.3). As PR expression in neonatal MPN is exclusively dependent on actions of estrogens at ERα, these findings suggest that BPA exerts estrogenic activity at ERα in developing MPN. These results are consistent with *in vitro* findings demonstrating that BPA induced PR mRNA and protein in rat pituitary GH3 cells, an effect that was blocked by ICI.[53] The induction of PR expression by direct administration of BPA to neonates is

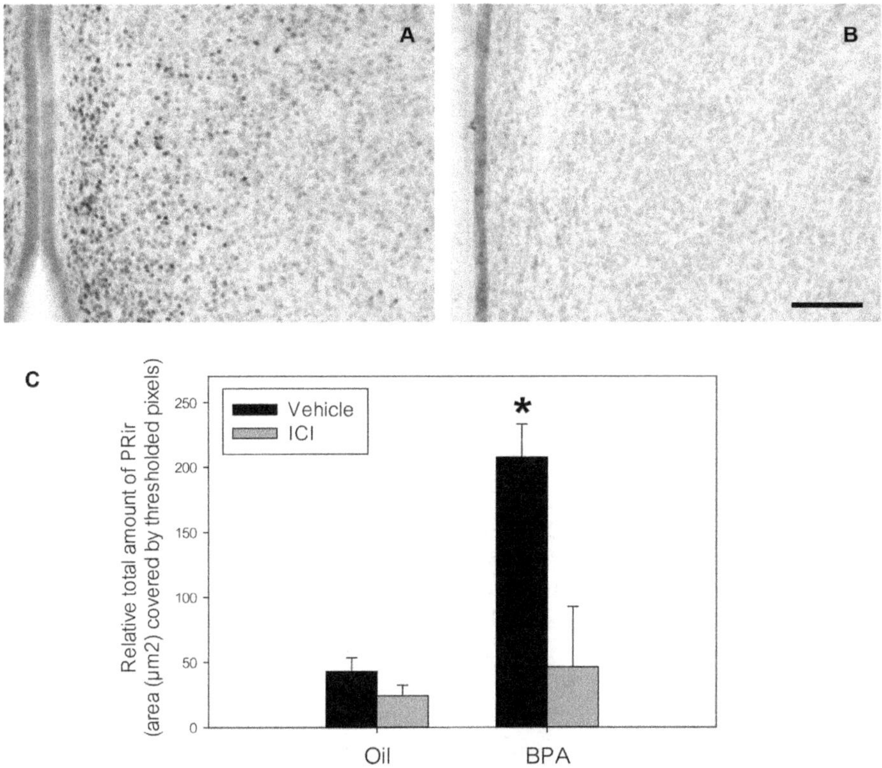

Figure 6.3 (A, B) Progesterone receptor immunoreactivity (PRir) in the medial preoptic nucleus of postnatal day 5 females who received (A) bisphenol A (BPA) and vehicle or (B) BPA and the estrogen receptor antagonist ICI 182,780 (ICI). Scale bar = 100 µm. (C) The relative total amount of PRir in the MPN of postnatal day 5 females who received BPA or oil vehicle (Oil) and ICI or vehicle. *: significantly different from all other groups ($p < 0.001$). Reproduced from ref. 49 with permission from Elsevier, Copyright 2020.

also consistent with the idea that fetal PR expression following maternally ingested BPA was a direct effect of BPA within the fetal brain and not simply the indirect result of, for example, increased transfer of endogenous estradiol from mother to fetus. Further, these results demonstrate that even low doses of BPA can induce PR expression in the neonatal female MPN, expression that normally would not occur until the second postnatal week, an outcome that could potentially masculinize the female MPN and disrupt the normal process of sexual differentiation.[54]

6.5 Conclusions

The induction of PR expression in the female fetal brain by maternal ingestion of a relatively low dose of BPA is consistent with previous literature demonstrating estrogenic-like effects of BPA despite its lower affinity for ERα and ERβ compared to estradiol. The absence of an effect on PR expression by BPA in the presence of an ER antagonist[49] and the almost exclusive dependence of PR expression in the perinatal MPN on ERα, and not ERβ activity,[47,48] suggest that BPA is transferred from the mother to the fetal brain where it exerts its effects directly on ERα. These findings offer several contributions. First, they introduce PR expression in the developing brain as a sensitive bio-indicator of ERα activation by maternal BPA. Second, they implicate progesterone receptor expression as one mechanism by which BPA may influence brain development. Third, they add to the emerging evidence that BPA exposure can alter the mechanisms of sexual differentiation of the immature brain. Our findings are consistent with previous findings demonstrating that BPA induced PR expression in the uterus, and the effect was abolished by the administration of ICI 182,780[6] and a single injection of BPA (10 mg) increased PR mRNA in the preoptic area of adult ovariectomized female rats only 24 hours after administration.[55]

BPA may exert masculinizing effects on the developing MPN through the induction of PR. Preliminary RNA sequencing studies from our lab suggest that PR activity in the first week of life is critical for the up- or down-regulation of 193 genes in neonatal male MPN.[64] These genes are presumably differentially expressed in the absence of PR expression in females. BPA exposure during this critical developmental window in females, even at relatively low doses, may induce PR-dependent changes in gene transcription that could alter the female-typical development of the MPN. Moreover, the potential exists for a feed-forward mechanism in the induction of PR expression by BPA in the perinatal female MPN. BPA exposure increases Esr1 (ERα) expression in some regions of the postnatal brain,[56] although findings are less clear for the neonatal preoptic area.[57] In our model, BPA-induced increases in levels of ERα in the perinatal female MPN would result in an amplification of BPA effects on ERα-dependent PR expression. In this way, BPA exposure would presumably make MPN neurons even more sensitive to low doses of maternal BPA. This may contribute to the sensitivity of this PR/MPN model system.

Numerous endocrine disruptors and other environmental contaminants, including BPA, exert non-monotonic dosing outcomes. In effect, lower doses of BPA often produce a physiological response while higher doses fail to do so,[3,50,51] contributing to the existing discrepancies in the literature on the severity of adverse effects of BPA.[12] Findings from our "PR in MPN" model are consistent with these findings. The lower dose of BPA (10 $\mu g\,kg^{-1}$ per day) induced PR in the MPN, while the higher dose (50 $\mu g\,kg^{-1}$ per day) did not. In a similar finding on steroid receptor expression, Rebuli *et al.*[58] reported that oral gavage of BPA to pregnant mothers followed by administration to offspring from birth to postnatal day 21 decreased ERβ mRNA levels in the preoptic area at the lowest dose (2.5 $\mu g\,kg^{-1}$), but higher doses (25, 260, and 2700 $\mu g\,kg^{-1}$) resulted in no change in ERβ expression. The "PR in MPN" model lends itself well as a sensitive outcome measure for a full dose–response study.

6.6 Future Directions

We suggest that PR expression in the perinatal MPN provides an innovative and sensitive developmental "bioassay" for the estrogenic effects of BPA at ERα in the developing brain. The potential for differential effects of BPA in perinatal females and males due to agonist and antagonist actions of BPA in the absence or presence of estradiol add to the value of the model. Furthermore, this model should be particularly useful in the investigation of the estrogenic effects of BPF and BPS, as "safe" alternatives of BPA. Little is understood about the physiological effects of these analogs, particularly in developing brain, despite their common use in the manufacture of "BPA-free" consumer products.[59,60] Pregnant women and their fetuses, as well as infants and children, may continue to be exposed to these compounds, the developmental effects of which are only beginning to come to light.[61–63]

References

1. A. M. Calafat, X. Ye, L.-Y. Wong, J. A. Reidy and L. L. Needham, Exposure of the US population to bisphenol A and 4-tertiary-octylphenol: 2003–2004, *Environ. Health Perspect.*, 2008, **116**, 39–44.
2. J. S. LaKind and D. Q. Naiman, Temporal trends in bisphenol A exposure in the United States from 2003–2012 and factors associated with BPA exposure: Spot samples and urine dilution complicate data interpretation, *Environ. Res.*, 2015, **142**, 84–95.
3. L. N. Vandenberg, I. Chahoud, J. J. Heindel, V. Padmanabhan, F. J. Paumgartten and G. Schoenfelder, Urinary, circulating, and tissue biomonitoring studies indicate widespread exposure to bisphenol A, *Environ. Health Perspect.*, 2010, **118**, 1055–1070.
4. A. Mendy, P. M. Salo, J. Wilkerson, L. Feinstein, K. K. Ferguson, M. B. Fessler, P. S. Thorne and D. C. Zeldin, Association of urinary levels of bisphenols F and S used as bisphenol A substitutes with asthma and hay fever outcomes, *Environ. Res.*, 2020, **183**, 108944.
5. Y. B. Wetherill, B. T. Akingbemi, J. Kanno, J. A. McLachlan, A. Nadal, C. Sonnenschein, C. S. Watson, R. T. Zoeller and S. M. Belcher, In vitro molecular mechanisms of bisphenol A action, *Reprod. Toxicol.*, 2007, **24**, 178–198.
6. J. C. Gould, L. S. Leonard, S. C. Maness, B. L. Wagner, K. Conner, T. Zacharewski, S. Safe, D. P. McDonnell and K. W. Gaido, Bisphenol A interacts with the estrogen receptor α in a distinct manner from estradiol, *Mol. Cell. Endocrinol.*, 1998, **142**, 203–214.
7. H. Hiroi, O. Tsutsumi, M. Momoeda, Y. Takai, Y. Osuga and Y. Taketani, Differential interactions of bisphenol A and 17β-estradiol with estrogen receptor α (ERα) and ERβ, *Endocr. J.*, 1999, **46**, 773–778.
8. T. Kurosawa, H. Hiroi, O. Tsutsumi, T. Ishikawa, Y. Osuga, T. Fujiwara, S. Inoue, M. Muramatsu, M. Momoeda and Y. Taketani, The activity of

bisphenol A depends on both the estrogen receptor subtype and the cell type, *Endocr. J.*, 2002, **49**, 465–471.
9. S. Hockfield and R. McKay, Identification of major cell classes in the developing mammalian nervous system, *J. Neurosci.*, 1985, **5**, 3310–3328.
10. H. MacKay and A. Abizaid, A plurality of molecular targets: The receptor ecosystem for bisphenol-A (BPA), *Horm. Behav.*, 2017, **101**, 59–67.
11. H. B. Patisaul, Achieving CLARITY on bisphenol A, brain and behaviour, *J. Neuroendocrinol.*, 2020, **32**, e12730.
12. F. S. Vom Saal and L. N. Vandenberg, Update on the health effects of bisphenol A: Overwhelming evidence of harm, *Endocrinology*, 2021, **162**, bqaa171.
13. D. Chen, K. Kannan, H. Tan, Z. Zheng, Y.-L. Feng, Y. Wu and M. Widelka, Bisphenol analogues other than BPA: environmental occurrence, human exposure, and toxicity: a review, *Environ. Sci. Technol.*, 2016, **50**, 5438–5453.
14. Y. Ikezuki, O. Tsutsumi, Y. Takai, Y. Kamei and Y. Taketani, Determination of bisphenol A concentrations in human biological fluids reveals significant early prenatal exposure, *Hum. Reprod.*, 2002, **17**, 2839–2841.
15. G. Schönfelder, W. Wittfoht, H. Hopp, C. E. Talsness, M. Paul and I. Chahoud, Parent bisphenol A accumulation in the human maternal-fetal-placental unit, *Environ. Health Perspect.*, 2002, **110**, A703–A707.
16. J. Lee, K. Choi, J. Park, H.-B. Moon, G. Choi, J. J. Lee, E. Suh, H.-J. Kim, S.-H. Eun and G.-H. Kim, Bisphenol A distribution in serum, urine, placenta, breast milk, and umbilical cord serum in a birth panel of mother–neonate pairs, *Sci. Total Environ.*, 2018, **626**, 1494–1501.
17. M. Zbucka-Krętowska, U. Łazarek, W. Miltyk, I. Sidorkiewicz, P. Pierzyński, R. Milewski, S. Wołczyński and J. Czerniecki, Simultaneous analysis of bisphenol A fractions in maternal and fetal compartments in early second trimester of pregnancy, *J. Perinat. Med.*, 2019, **47**, 765–770.
18. B. Balakrishnan, K. Henare, E. B. Thorstensen, A. P. Ponnampalam and M. D. Mitchell, Transfer of bisphenol A across the human placenta, *Am. J. Obstet. Gynecol.*, 2010, **202**(393), e391–e397.
19. X. Ye, Z. Kuklenyik, L. L. Needham and A. M. Calafat, Measuring environmental phenols and chlorinated organic chemicals in breast milk using automated on-line column-switching–high performance liquid chromatography–isotope dilution tandem mass spectrometry, *J. Chromatogr.*, 2006, **831**, 110–115.
20. Y. Sun, M. Irie, N. Kishikawa, M. Wada, N. Kuroda and K. Nakashima, Determination of bisphenol A in human breast milk by HPLC with column-switching and fluorescence detection, *Biomed. Chromatogr.*, 2004, **18**, 501–507.
21. R. Kuruto-Niwa, Y. Tateoka, Y. Usuki and R. Nozawa, Measurement of bisphenol A concentrations in human colostrum, *Chemosphere*, 2007, **66**, 1160–1164.

22. A. C. Gore, K. Krishnan and M. P. Reilly, Endocrine-disrupting chemicals: Effects on neuroendocrine systems and the neurobiology of social behavior, *Horm. Behav.*, 2019, **111**, 7–22.
23. D. Nesan and D. M. Kurrasch, Gestational exposure to common endocrine disrupting chemicals and their impact on neurodevelopment and behavior, *Annu. Rev. Physiol.*, 2020, **82**, 177–202.
24. K. Itoh, T. Yaoi and S. Fushiki, Bisphenol A, an endocrine-disrupting chemical, and brain development, *Neuropathology*, 2012, **32**, 447–457.
25. K. Nakamura, K. Itoh, T. Sugimoto and S. Fushiki, Prenatal exposure to bisphenol A affects adult murine neocortical structure, *Neurosci. Lett.*, 2007, **420**, 100–105.
26. K. Nakamura, K. Itoh, T. Yaoi, Y. Fujiwara, T. Sugimoto and S. Fushiki, Murine neocortical histogenesis is perturbed by prenatal exposure to low doses of bisphenol A, *J. Neurosci. Res.*, 2006, **84**, 1197–1205.
27. M. Komada, T. Nagao and N. Kagawa, Prenatal and postnatal bisphenol A exposure inhibits postnatal neurogenesis in the hippocampal dentate gyrus, *J. Toxicol. Sci.*, 2020, **45**, 639–650.
28. M. Yokosuka, R. Ohtani-Kaneko, K. Yamashita, D. Muraoka, Y. Kuroda and C. Watanabe, Estrogen and environmental estrogenic chemicals exert developmental effects on rat hypothalamic neurons and glias, *Toxicol. In Vitro*, 2008, **22**, 1–9.
29. H. Shikimi, H. Sakamoto, Y. Mezaki, K. Ukena and K. Tsutsui, Dendritic growth in response to environmental estrogens in the developing Purkinje cell in rats, *Neurosci. Lett.*, 2004, **364**, 114–118.
30. N. J. MacLusky, T. Hajszan and C. Leranth, The environmental estrogen bisphenol A inhibits estradiol-induced hippocampal synaptogenesis, *Environ. Health Perspect.*, 2005, **113**, 675–679.
31. Z. Wang, M. H. Alderman, C. Asgari and H. S. Taylor, Fetal bisphenol-A induced changes in murine behavior and brain gene expression persisted in adult-aged offspring, *Endocrinology*, 2020, **161**, bqaa164.
32. J. T. Wolstenholme, E. F. Rissman and J. J. Connelly, The role of Bisphenol A in shaping the brain, epigenome and behavior, *Horm. Behav.*, 2011, **59**, 296–305.
33. M. León-Olea, C. J. Martyniuk, E. F. Orlando, M. A. Ottinger, C. S. Rosenfeld, J. T. Wolstenholme and V. L. Trudeau, Current concepts in neuroendocrine disruption, *Gen. Comp. Endocrinol.*, 2014, **203**, 158–173.
34. K. Kawai, T. Nozaki, H. Nishikata, S. Aou, M. Takii and C. Kubo, Aggressive behavior and serum testosterone concentration during the maturation process of male mice: the effects of fetal exposure to bisphenol A, *Environ. Health Perspect.*, 2003, **111**, 175–178.
35. J. M. Braun, K. Yolton, K. N. Dietrich, R. Hornung, X. Ye, A. M. Calafat and B. P. Lanphear, Prenatal bisphenol A exposure and early childhood behavior, *Environ. Health Perspect.*, 2009, **117**, 1945–1952.
36. S. F. Evans, R. W. Kobrosly, E. S. Barrett, S. W. Thurston, A. M. Calafat, B. Weiss, R. Stahlhut, K. Yolton and S. H. Swan, Prenatal bisphenol A

exposure and maternally reported behavior in boys and girls, *Neurotoxicology*, 2014, **45**, 91–99.
37. H. B. Patisaul, A. E. Fortino and E. K. Polston, Neonatal genistein or bisphenol-A exposure alters sexual differentiation of the AVPV, *Neurotoxicol. Teratol.*, 2006, **28**, 111–118.
38. H. B. Patisaul, Effects of environmental endocrine disruptors and phytoestrogens on the kisspeptin system, *Adv. Exp. Med. Biol.*, 2013, 455–479.
39. H. B. Patisaul, A. E. Fortino and E. K. Polston, Differential disruption of nuclear volume and neuronal phenotype in the preoptic area by neonatal exposure to genistein and bisphenol-A, *Neurotoxicology*, 2007, **28**, 1–12.
40. H. B. Patisaul and E. K. Polston, Influence of endocrine active compounds on the developing rodent brain, *Brain Res. Rev.*, 2008, **57**, 352–362.
41. C. K. Wagner, A. Y. Nakayama and G. J. De Vries, Potential role of maternal progesterone in the sexual differentiation of the brain, *Endocrinology*, 1998, **139**, 3658–3661.
42. B. S. Katzenellenbogen, I. Choi, R. Delage-Mourroux, T. R. Ediger, P. G. Martini, M. Montano, J. Sun, K. Weis and J. A. Katzenellenbogen, Molecular mechanisms of estrogen action: selective ligands and receptor pharmacology, *J. Steroid Biochem. Mol. Biol.*, 2000, **74**, 279–285.
43. W. L. Kraus, M. M. Montano and B. S. Katzenellenbogen, Identification of multiple, widely spaced estrogen-responsive regions in the rat progesterone receptor gene, *Mol. Endocrinol.*, 1994, **8**, 952–969.
44. P. S. Quadros, A. Y. N. Goldstein, G. J. De Vries and C. K. Wagner, Regulation of sex differences in progesterone receptor expression in the medial preoptic nucleus of postnatal rats, *J. Neuroendocrinol.*, 2002, **14**, 761–767.
45. P. S. Quadros, J. L. Pfau, A. Y. N. Goldstein, G. J. De Vries and C. K. Wagner, Sex differences in progesterone receptor expression: A potential mechanism for estradiol-mediated sexual differentiation, *Endocrinology*, 2002, **143**, 3727–3739.
46. K. L. Gonzales, P. Quadros-Mennella, M. J. Tetel and C. K. Wagner, Anatomically-specific actions of oestrogen receptor in the developing female rat brain: effects of oestradiol and selective oestrogen receptor modulators on progestin receptor expression, *J. Neuroendocrinol.*, 2012, **24**, 285–291.
47. K. Gonzales, M. Tetel and C. Wagner, Estrogen receptor β (ERβ) modulates ERα responses to estrogens in the developing rat ventromedial nucleus of the hypothalamus, *Endocrinology*, 2008, **149**, 4615–4621.
48. C. K. Wagner, J. L. Pfau, G. J. De Vries and I. J. Merchenthaler, Sex differences in progesterone receptor immunoreactivity in neonatal mouse brain depend on estrogen receptor expression, *J. Neurobiol.*, 2001, **47**, 176–182.

49. A. Fahrenkopf and C. K. Wagner, Bisphenol A (BPA) induces progesterone receptor expression in an estrogen receptor α-dependent manner in perinatal brain, *Neurotoxicol. Teratol.*, 2020, **78**, 106864.
50. F. Lagarde, C. Beausoleil, S. M. Belcher, L. P. Belzunces, C. Emond, M. Guerbet and C. Rousselle, Non-monotonic dose-response relationships and endocrine disruptors: a qualitative method of assessment, *Environ. Health*, 2015, **14**, 13–28.
51. L. N. Vandenberg, I. Chahoud, J. J. Heindel, V. Padmanabhan, F. J. Paumgartten and G. Schoenfelder, Urinary, circulating, and tissue biomonitoring studies indicate widespread exposure to bisphenol A, *Environ. Health Perspect.*, 2010, **118**, 1055–1070.
52. L. N. Vandenberg, R. R. Gerona, K. Kannan, J. A. Taylor, R. B. van Breemen, C. A. Dickenson, C. Liao, Y. Yuan, R. R. Newbold and V. Padmanabhan, A round robin approach to the analysis of bisphenol A (BPA) in human blood samples, *Environ. Health*, 2014, **13**, 25.
53. S. Kim, E. Jung, B. An, I. Hwang, T. Vo, S. Kim, S. Lee, K. Choi and E. Jeung, Additional effects of bisphenol A and paraben on the induction of calbindin-D (9K) and progesterone receptor via an estrogen receptor pathway in rat pituitary GH3 cells, *J. Physiol. Pharmacol.*, 2012, **63**, 445–455.
54. P. S. Quadros, V. Lopez, G. J. De Vries, W. C. J. Chung and C. K. Wagner, Progesterone receptors and the sexual differentiation of the medial preoptic nucleus, *J. Neurobiol.*, 2002, **51**, 24–32.
55. T. Funabashi, M. Kawaguchi and F. Kimura, The endocrine disrupters butyl benzyl phthalate and bisphenol A increase the expression of progesterone receptor messenger ribonucleic acid in the preoptic area of adult ovariectomized rats, *Neuroendocrinology*, 2001, **74**, 77–81.
56. L. M. Wise, D. Hernández-Saavedra, S. M. Boas, Y.-X. Pan and J. M. Juraska, Perinatal high-fat diet and bisphenol A: Effects on behavior and gene expression in the medial prefrontal cortex, *Dev. Neurosci.*, 2019, **41**, 1–16.
57. J. Cao and H. B. Patisaul, Sex-specific expression of estrogen receptors α and β and Kiss1 in the postnatal rat amygdala, *J. Comp. Neurol.*, 2013, **521**, 465–478.
58. M. E. Rebuli, J. Cao, E. Sluzas, K. B. Delclos, L. Camacho, S. M. Lewis, M. M. Vanlandingham and H. B. Patisaul, Investigation of the effects of subchronic low dose oral exposure to bisphenol A (BPA) and ethinyl estradiol (EE) on estrogen receptor expression in the juvenile and adult female rat hypothalamus, *Toxicol. Sci.*, 2014, **140**, 190–203.
59. J. R. Rochester and A. L. Bolden, Bisphenol S and F: a systematic review and comparison of the hormonal activity of bisphenol A substitutes, *Environ. Health Perspect.*, 2015, **123**, 643–650.
60. C. Liao and K. Kannan, Concentrations and profiles of bisphenol A and other bisphenol analogues in foodstuffs from the United States and their implications for human exposure, *J. Agric. Food Chem.*, 2013, **61**, 4655–4662.

61. J. Gingrich, D. Filipovic, R. Conolly, S. Bhattacharya and A. Veiga-Lopez, Pregnancy-specific physiologically-based toxicokinetic models for bisphenol A and bisphenol S, *Environ. Int.*, 2021, **147**, 106301.
62. M. Naderi and R. W. Kwong, A comprehensive review of the neurobehavioral effects of bisphenol S and the mechanisms of action: New insights from in vitro and in vivo models, *Environ. Int.*, 2020, **145**, 106078.
63. K. Pelch, J. A. Wignall, A. E. Goldstone, P. K. Ross, R. B. Blain, A. J. Shapiro, S. D. Holmgren, J.-H. Hsieh, D. Svoboda and S. S. Auerbach, A scoping review of the health and toxicological activity of bisphenol A (BPA) structural analogues and functional alternatives, *Toxicology*, 2019, **424**, 152235.
64. Lalitsasivimol *et al.*, unpublished.

CHAPTER 7

Pleiotropic Effects of Bisphenol A Alter Milk Synthesis and Production

GABRIELA A. ALTAMIRANO,[a,b] AYELEN L. GOMEZ[a,b] AND LAURA KASS*[a,b]

[a] Instituto de Salud y Ambiente del Litoral (ISAL, UNL-CONICET), Facultad de Bioquímica y Ciencias Biológicas, Universidad Nacional del Litoral, Ciudad Universitaria, Paraje El Pozo, Casilla de Correo 242, (3000) Santa Fe, Argentina; [b] Cátedra de Patología Humana, Facultad de Bioquímica y Ciencias Biológicas, Universidad Nacional del Litoral, Santa Fe, Argentina
*Email: lkass@fbcb.unl.edu.ar

7.1 Introduction

Milk is a nutrient-rich liquid food produced and secreted by the female mammary gland in all mammals, which is considered the primary nutrition source for the newborn. Milk composition includes a variety of key nutrients such as milk proteins, milk fat globules (MFGs), carbohydrates, vitamins, minerals, and growth factors that are suspended or dissolved in water.[1,2] Many of these components are transferred directly from the maternal circulation, whereas others are synthesized *de novo* by mammary epithelial cells.[1] Thus, the correct differentiation of the mammary alveolar cells is necessary for a suitable synthesis and secretion of milk components. Milk composition also depends on the mother's stage of lactation, diet, age, and her exposure to environmental stressors, such as endocrine-disrupting

Issues in Toxicology No. 43
Bisphenol A: A Multi-modal Endocrine Disruptor
Edited by Natalie R. Gassman

Published by the Royal Society of Chemistry, www.rsc.org

chemicals (EDCs), during gestation and lactation.[3,4] In recent decades, the plastic monomer and plasticizer bisphenol A (BPA) has attracted special attention because of its proven estrogenic activity that causes adverse effects on animal and human health.[5,6] In the female mammary gland, perinatal exposure to BPA modifies the mammary histo-architecture, inducing an alteration in its maturation rate, a delay in lumen formation, and an increase in ductal growth. Additionally, perinatal exposure to BPA modifies the ability of the mammary gland to respond to a secondary estrogenic exposure, increasing its susceptibility to the development of mammary lesions in rodents.[7–11] Although all this evidence suggests that perinatal exposure to BPA may predispose individuals to breast cancer later in life, it could also indicate that the functional differentiation of the mammary gland, and consequently, the synthesis and composition of milk, may be impaired. Furthermore, BPA is transferred from blood to the breast by passive diffusion, and, due to its lipophilic properties, it is observed at high concentrations in human milk.[12,13] In this chapter, we describe and discuss the effects of direct and perinatal exposure to BPA on functional differentiation of the rodent mammary gland, including how it affects milk production and composition, mammary histo-architecture, and the mammary endocrine microenvironment, as well as the multiple alternative mechanisms of action of BPA that are involved, in addition to the classical estrogenic pathway.

7.2 Milk Production and its Nutritional Quality

The synthesis and secretion of milk components depend not only on the correct differentiation of the mammary gland during pregnancy, but also on maternal adaptations during lactation, including milk ejection, maternal behavior, and a reduced stress response.[14] Hormones such as progesterone, estradiol, and prolactin play an important role during pregnancy and the prepartum period, activating the organization of the neuroendocrine substrates responsible for maternal behavior initiation.[15] Therefore, the effects of prenatal or perinatal exposure to EDCs on the hormonal environment of the dams could be reflected in changes in maternal behavior during lactation. Furthermore, maternal stress could influence the milk ejection reflex by interfering with the release of oxytocin. Consequently, the incomplete removal of milk from the mammary gland could lead to a down-regulation of milk synthesis.[16] Taken together, changes in maternal behavior induced by perinatal exposure to EDC could interfere with milk production and its nutritional quality in F1 dams. These changes are significant because they could also compromise the development and growth of the F2 offspring. To investigate the influence of perinatal exposure to BPA on milk quality, pregnant Wistar rats (F0) were exposed *via* drinking water to vehicle or 0.6 or 52 $\mu g\,kg^{-1}$ BPA per day (BPA0.6 or BPA52, respectively) from gestation day 9 (GD9) until weaning on postnatal day 21 (PND21). F1 female offspring were transferred to an EDC-free environment, and when they reached adulthood, they were bred with unexposed males. Then, F1 dams were randomly distributed to evaluate the influence of perinatal exposure to BPA on milk production, synthesis, and composition during lactation, the histo-architecture of the mammary gland during its secretory activation, and the bodyweight gain of their F2 offspring.[17,18] In the following sections, we discuss the results obtained with this experimental model.

7.2.1 Perinatal Exposure to BPA Affects Maternal Behavior, Milk Production, and the Development of the Lactating Mammary Gland

Perinatal exposure to BPA has been shown to induce a negative balance of milk production in F1 dams without apparent alterations in nursing (F1) or suckling (F2) behavior during most of the first 13 days of lactation.[18] However, exposure to BPA can decrease the maternal behavior of F1 dams, such as the time that dams spend in the nest and lick and/or groom their F2 offspring during the first half of lactation.[19–21] Palanza *et al.*[15] postulated that these maternal behavior changes may be due to the direct effect of BPA on the neuroendocrine substrates underlying the initiation of maternal behavior. Exposure to BPA could also induce epigenetic alterations in F1 dams that are reflected in maternal behavior.[19,22] A recent study has demonstrated that exposure during development to a mixture of EDCs, including BPA, induces a multigenerational reduction of maternal behavior due to epigenetic reprogramming of the hypothalamus.[23] In this study, the decrease in maternal behavior was associated with a lower expression of key genes involved in the dopaminergic control of maternal behavior and epigenetic alterations of the dopaminergic system in the hypothalamus of F1 dams.[23] Furthermore, perinatal exposure to BPA is thought to act as an agonist of the glucocorticoid receptor (GR) on the hypothalamic–pituitary–adrenal axis in rodents.[6,24] It is known that high levels of corticosterone in the postpartum stage and/or stressful situations could influence the emotional response of the F1 offspring in adulthood,[25] increasing their vulnerability to situations of high emotional demand, such as pregnancy and lactation. The stressful situation induced by the separation of F1 mothers from their F2 offspring in the lactational challenge could cause a temporary reduction in milk ejection reflected in decreased milk production in BPA-exposed F1 dams (Figure 7.1).[18] Caparros-Gonzalez *et al.*[26] reported that maternal stress and high cortisol levels during the last stage of pregnancy delays the onset of the secretory activation of the mammary gland. Perinatal exposure to BPA has been shown to delay maturation of mammary alveoli in the rat (Figure 7.1). In F1 females exposed to BPA, the mammary alveoli were less distended compared to controls at the end of gestation and early lactation, whereas they were re-established at mid-lactation (lactation day (LD)10).[17,18] This delay in the onset of mammary secretory activation could be associated with maternal stress indirectly induced by exposure to BPA. In addition, increasing evidence suggests that EDCs may affect the duration of breastfeeding by impacting the development and functional differentiation of the mammary gland.[27] In a limited cohort of Mexican women, higher maternal urinary BPA levels were associated with a reduced duration of breastfeeding.[28] The decrease in milk production and the delay in the onset of mammary secretory activation, induced by perinatal exposure to BPA, could be in part the reason why mothers decided to stop breastfeeding and use formula milk.

7.2.2 Perinatal Exposure to BPA Modifies Milk Quality and Weight in F2 Offspring

The negative balance of milk production and the delay of mammary secretory activation observed in BPA-exposed F1 dams are probably associated with changes in the synthesis and/or secretion of the different components of the milk (Figure 7.1). In milk samples collected on LD10, the concentration of total lipids, triglycerides, phospholipids in milk and the fatty acid (FA) profile of milk was altered by perinatal exposure to BPA.[17] The protein expression of alpha-lactalbumin (LALBA) was similar between groups, whereas beta-casein (CSN2) was decreased in BPA-exposed F1 dams on LD14.[18] These changes in

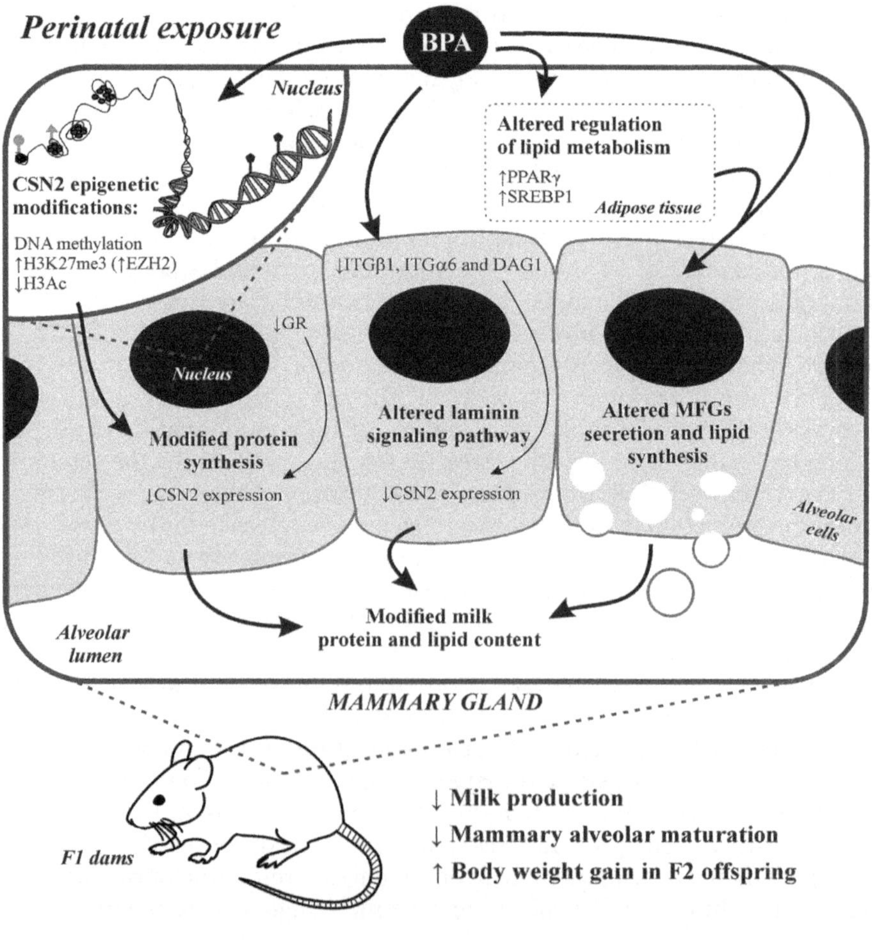

the quality of the milk may alter the growth and development of F2 offspring. The high lipid content in milk found in BPA-exposed F1 dams increased bodyweight gain in F2 offspring between LD5 and LD9 (Figure 7.1).[17] Thus, this difference in bodyweight gain in F2 offspring was associated with changes in milk fat composition rather than milk protein on LD10.[17]

Several studies have reported that prenatal and/or early-life exposure to EDCs can strongly influence an increase in bodyweight later in life and modify adipogenesis by trans-generational effects.[29,30] Interestingly, Boudalia *et al.*[19] have registered the bodyweight of F2 offspring of BPA-exposed F1 dams from weaning to adulthood. In this study, the authors showed an increase in bodyweight of F2 offspring from PND76 to PND94 compared with controls. They presumed that the increase was due to a modification of the central (hypothalamus) or the peripheral (adipose tissue) control of energy homeostasis.[19] In addition, it is known that changes in the offspring growth due to over-nutrition during the early postnatal period may predispose to the development of metabolic disease.[31] Considering that this early nutritional programming is related to milk composition, milk quality changes may have long-lasting programming/deprogramming metabolic effects in the offspring.[31] Taken together, the effects of perinatal exposure to BPA on milk production and quality plus the increase of bodyweight gain of F2 offspring

Figure 7.1 Effects of perinatal exposure to bisphenol A (BPA) on milk synthesis in the rat mammary gland. In BPA-exposed F1 dams, a decline in milk production was observed during the first half of lactation. This change was accompanied by a delay in the mammary alveolar maturation during mammary secretory activation, and changes in both milk lipid and protein content at mid-lactation. These modifications in milk quality were reflected in an increased bodyweight gain in F2 offspring. Perinatal exposure to BPA altered lipid metabolism regulation, increasing PPARγ and SREBP1 mRNA expression at mid-lactation. This increase altered both milk fat globule (MFG) secretion and lipid synthesis in the mammary gland. Consequently, a modified milk fatty acid profile was observed in F1 dams. Among milk proteins, perinatal exposure to BPA decreased CSN2 expression during mammary secretory activation by altering different molecular mechanisms that seem to act synergistically. Perinatal exposure to BPA decreased the mRNA expression of glucocorticoid receptor (GR) and laminin-binding receptors (ITGα6, ITGβ1, and DAG1), leading possibly to a transient Stat5 activation not sufficient to activate CSN2 gene expression. Perinatal exposure to BPA also increased EZH2 mRNA expression and increased H3K27me3 levels in regulatory regions of CSN2 gene. A decreased level of H3Ac was also observed at the CSN2 promoter and enhancer. Despite the decreased level of DNA methylation observed in the CSN2 gene, these histone modifications resulted in the chromatin structure's compaction and, consequently, a decreased transcription of the CSN2 gene. CSN2: beta-casein; DAG1: dystroglycan 1; EZH2: enhancer of zeste homolog 2; F1: dams perinatally exposed to bisphenol A; F2: F1 offspring; H3Ac: acetyl histone 3; H3K27me3: trimethyl-histone H3 (Lys 27); ITGα6: integrin alpha 6; ITGβ1: integrin beta 1; PPARγ: peroxisome proliferator-activated receptor gamma; SREBP1: sterol regulatory element-binding protein 1; Stat5: signal transducer and activator of transcription 5.

observed during the first half of lactation are likely to be associated with metabolic diseases such as obesity in adulthood.

7.3 Milk Lipid Synthesis

The highest proportion of lipids in the milk are triglycerides (<98%),[32] which, due to their high energy content, provide most of the calories required by the growing newborn in most species.[33] Milk lipids are considered the main source of essential FAs necessary for the synthesis of the neonatal membrane. They are also substrates for the synthesis of eicosanoids and other bioactive lipid signaling molecules. Finally, they provide a mechanism for the transfer of fat-soluble vitamins to infants.[34] The lipid content of milk differs between species and varies through lactation, changing through the lactation stages.[2] Considering that perinatal exposure to BPA delays mammary alveolar maturation at secretory activation and alters the quality of the milk lipid fraction in F1 dams at mid-lactation,[17,18] the synthesis of triglycerides and secretion of MFGs could be disrupted during lactation.

7.3.1 Perinatal Exposure to BPA Alters Milk Lipid Secretion and Accumulation

In mammary epithelial cells, the secretion of milk lipids is a specialized process that involves specific molecular interactions between the cytoplasmic lipid droplet (CLD) and the apical plasma membrane elements to produce membrane bound-secreted products called MFGs that are released into the alveolar lumen.[35] It has been described that MFG secretion occurs through the formation of a tripartite complex between adipophilin (ADPH) located on the surface of CLDs, the transmembrane protein butyrophilin (BTN) and the cytoplasmic enzyme xanthine oxidoreductase (XOR).[35] In this process, BTN functions as an integral receptor for CLD by recognizing ADPH on its surface. BTN is responsible for initiating the budding process of CLD in the apical cell membrane through complex formation with XOR.[35] Thus, deficiencies in XOR and/or BTN, or the interference with their interactions with ADPH, have been predicted to perturb the secretion of MFGs.[35,36] The CLD accumulation pattern suggests that milk lipid formation involves sequentially regulated lipogenic, transport, and secretion patterns that are functionally linked to mammary alveolar maturation.[34,37] Perinatal exposure to BPA altered the mRNA expression of BTN, XOR and ADPH during late pregnancy and early lactation, indicating that the assembly and/or secretion of MFGs could be compromised.[17] On GD21 and LD2, the mRNA levels of ADPH, BTN, and XOR were decreased in the mammary gland of BPA-exposed F1 dams, suggesting that the MFG secretion could be reduced during the mammary secretory activation.[17] Furthermore, on LD2, the MFG size was decreased in BPA groups compared with control dams.[17] These results are in accordance with recent work in which the authors demonstrated that knockout of the BTN gene decreased both the content and diameter of CLD in bovine mammary epithelial cells.[38] In this study, BTN gene knockout decreased the nuclear receptor peroxisome proliferator-activated receptor gamma (PPARγ).[38] PPARγ mediates lipid accumulation by inducing lipogenic genes such as ADPH and FA synthase (FAS) enzyme in mammary cells.[38] In BPA-exposed F1 dams, the lower expression of BTN mRNA and smaller size of the MFGs suggest that the activity of PPARγ and, therefore, the synthesis of FA could be decreased at early lactation. Another recent study demonstrated that a high level of serum progesterone was associated with a decreased MFG size in mammary epithelial cells.[39] High serum progesterone prioritizes the use of circulating-derived long-chain FAs for phospholipid synthesis, resulting in the formation of smaller lipid droplets rather than triglyceride synthesis in mammary epithelial cells.[39] In this regard, an increased level of serum progesterone was observed in BPA-exposed F1 dams on GD21 compared with controls.[18] This finding could explain the reduced MFG size found later in the BPA-exposed F1 dams at early lactation. In contrast, at mid-lactation, both the MFG size and the mRNA levels of PPARγ and ADPH were increased in BPA-exposed F1 dams.[17] It is known that ADPH is the major protein component in CLD and regulates CLD size by preventing the access of lipases to the triglyceride core of CLDs.[40,41] Also, several studies have reported that ADPH expression correlates with lipid accumulation in milk-secreting cells during lactation.[43,44] Impaired CLD accumulation induced by ADPH inhibition has been reported to interfere with alveolar maturation by altering triglyceride homeostasis in mammary epithelial cells.[42] The lower ADPH mRNA expression found in BPA-exposed F1 dams during mammary secretory activation (GD21 and LD2) could be associated with the delay in the alveolar maturation observed in these dams.[17] In contrast, at mid-lactation, the alveolar maturation was restored, and an increase in both the size of the MFGs and ADPH mRNA expression were found in BPA-exposed F1 dams, indicating a greater secretion and accumulation of lipids in the mammary gland. Taken together, exposure to BPA appears to alter MFG homeostasis in the mammary gland by interference with BTN and ADPH expression. Furthermore, the MFG membrane is considered a bioactive ingredient contributing to the offspring's neurocognitive development by providing polar lipids such as gangliosides and sphingomyelin.[45] Alteration of the MFG membrane composition induced by low doses of BPA could negatively affect the development and function of the offspring brain.

7.3.2 Perinatal Exposure to BPA Modifies Milk Lipid Synthesis and Lipid Metabolism Regulation

Milk lipids, both triglycerides and phospholipids, are synthesized in the endoplasmic reticulum of the milk-secreting cells from precursors of FAs and glycerol.[46] In the mammary epithelial cells, FAs are either transported from the plasma *via* albumin or through hydrolysis of chylomicron triglycerides by lipoprotein lipase (LPL); or derived from *de novo* synthesis from the glucose that enters through reactions catalyzed mainly by the enzymes acetyl CoA carboxylase alpha (ACCα) and FAS.[32,47] Also, stearoyl-CoA desaturase 2 (SCD2) is a key enzyme in the cellular biosynthesis of monounsaturated FAs and is one of the most abundant SCD isoforms in the mouse mammary gland during lactation.[48] Therefore, FA synthesis is regulated both by the mammary gland's secretory activation and the availability of exogenous lipids.[32] The transcriptional levels of ACCα and LPL were decreased by perinatal exposure to BPA at the end of gestation, suggesting that FA synthesis is limited in the mammary gland during secretory activation.[17] In contrast, only SCD2 mRNA expression was increased in BPA52-exposed F1 dams on LD10.[17] As described earlier, the histo-architecture of the mammary gland was re-established, and the mRNA levels of most of the studied enzymes were not altered in BPA-exposed F1 dams at mid-lactation.[17] However, changes in the milk FAs profile were detected in F1 dams, demonstrating that FA synthesis could be altered by perinatal exposure to BPA (Figure 7.1). It has been reported that most of the short- and medium-chain FAs are derived from *de novo* FAs synthesis in the mammary epithelial cells, whereas most of the long-chain FAs are derived from the circulation as preformed FAs.[49] Perinatal exposure to BPA52 increased the levels of myristic acid, palmitic acid, and stearic acid in the milk, suggesting that *de novo* synthesis of FAs is altered in the lactating mammary gland.[17] Although the mRNA expression of FAS was similar between groups, the FAS activity could be increased in the F1 dams exposed to BPA52.[17] In contrast, the levels of short- and medium-chain FAs were reduced in BPA0.6-exposed F1 dams, whereas long-chain polyunsaturated FAs such as arachidonic acid and docosahexaenoic acid were increased in these dams.[17] This finding indicates that milk FA synthesis derived more from maternal circulation rather than *de novo* synthesis of FAs in the F1 dams exposed to BPA0.6.

As described earlier, PPARγ regulates the transcription of several genes involved in lipid metabolism in the mammary gland, either directly or through the activation of the transcription regulators, sterol regulatory-element binding protein (SREBP1) and liver X receptor alpha.[50] In turn, SREBP1 regulates *de novo* FA synthesis through ACCα, FAS, and SCD2.[51,52] Furthermore, Kang *et al.*[53] have shown that the activation of PPARγ signaling promotes the accumulation of lipids by increasing ADPH expression in goat mammary epithelial cells. The mRNA level expression of both PPARγ and SREBP1 were up-regulated in BPA52-exposed F1 dams at mid-lactation.[17] A recent review highlighted that one of the novel possible mechanisms of BPA action could be through the activation of PPARγ signaling, increasing lipid accumulation in adipocyte cells.[5] Interestingly, another recent *in vitro* study showed that exposure to BPA decreases DNA methylation of the PPARγ promoter, increasing mRNA expression of PPARγ and lipid accumulation in a pre-adipocyte cell line.[54] Perinatal exposure to BPA could directly or indirectly activate PPARγ signaling in the mammary gland, increasing both *de novo* FA synthesis and ADPH mRNA expression in BPA52-exposed F1 dams at mid-lactation (Figure 7.1).[17]

7.4 Milk Protein Synthesis

The main proteins synthesized and secreted in the mammary gland are caseins and the so-called whey proteins, including LALBA, whey acidic protein (WAP), beta-lactoglobulin and lactoferrin.[32] Caseins, which represent 80% of milk proteins, form insoluble micelles containing high concentrations of calcium and phosphates.[33] Among caseins, CSN2 is one of the major isoforms secreted during lactation.[55] Another important milk protein is LALBA, due to the abundance required for lactose synthesis.[1] In addition to their nutritional importance, milk proteins are considered functional differentiation markers in the mammary gland; within them, CSN2, LALBA, and WAP are the main ones.[56] Milk proteins genes are regulated principally by lactogenic hormones, which, together with the cooperation of local growth factors and cell–cell or cell–extracellular matrix (ECM) interactions, activate specific transcription factors, alter the organization of the cytoskeleton, and change the state of chromatin and nuclear structures.[57] Here, we describe how perinatal and direct exposure to BPA affects milk protein synthesis and secretion during the mammary gland's secretory activation through different hormonal pathways, cell–extracellular interactions, and epigenetic mechanisms.

7.4.1 Perinatal Exposure to BPA Modifies the Expression of Milk Proteins and Interferes with Lactogenic Hormone and Laminin Signaling Pathways in the Mammary Gland

As mentioned earlier, perinatal exposure to BPA delayed mammary alveolar maturation at secretory activation and decreased milk protein expression in milk at mid-lactation. In perinatally BPA-exposed animals, the mRNA expression of LALBA was only decreased on GD18, whereas mRNA and protein expression of CSN2 was decreased on GD18, GD21, and LD2 (Figure 7.1).[18,58] These results are in accordance with the expression of LALBA and CSN2 found in milk at mid-lactation, in which only CSN2 was reduced in BPA-exposed groups.[18]

It is known that the lactogenic hormone prolactin increases the transcription of milk protein genes by the activation of prolactin receptor (PRLR) and phosphorylation and dimerization of the signal transducer and activator of transcription 5 (Stat5) during the functional differentiation of the mammary gland.[32] In rats exposed perinatally to BPA, both the PRLR mRNA level and the percentage of positive nuclear cells for phosphorylated Stat5 a/b (pStat5a/b) were only decreased on GD18, with no significant differences on GD21 and LD2, indicating that the prolactin signaling pathway remains active during mammary secretory activation.[18,58] These findings suggest that the decrease in prolactin signaling reduces the levels of both LALBA and CSN2 expression on GD18, but is not responsible for the low expression of CSN2 during the secretory activation of the mammary gland.

Other lactogenic hormones that are critical during the mammary secretory activation are glucocorticoids, which co-operate synergistically with prolactin signaling to modulate CSN2 gene expression through the activation of the GR.[59] The ligand activation of GR induces the formation of a GR–pStat5a/b complex and the subsequent deviation to the Stat5-DNA binding site in both the CSN2 proximal promoter and distal enhancer.[60] At early lactation, the mRNA expression of GR was decreased in BPA52-exposed F1 dams, suggesting that it could be related to the low expression of CSN2 observed in these dams.[58]

In addition, the prolactin signaling pathway is positively influenced by specific integrins in response to signals from the ECM during lactation.[61] The basement membrane rich in laminin coupled to the mammary alveoli acts through integrin beta 1 (ITGβ1) to established apicobasal polarity and differentiation.[62] Laminin signaling is mediated by the integrin dimers α3/β1, α6/β1, and α6/β4 mammary epithelial receptors, which are necessary for both lobuloalveolar development and CSN2 gene expression induced by prolactin.[62,63] A recent review reported that the knockout of ITGβ1 gene impaired the secretory differentiation of mammary cells, decreased the activation of Stat5, and decreased the mRNA expression of milk proteins such as CSN2 and WAP.[59] Also, the laminin co-receptor dystroglycan 1 (DAG1) is required for the correct signaling through integrin dimers receptors to allow efficient CSN2 production in mammary epithelial cells.[64] Perinatal exposure to BPA decreased the expression of laminin-binding receptors (integrin alpha 6, ITGβ1, and DAG1), leading possibly to both a transient Stat5 activation that is not sufficient to activate CSN2 gene expression, as well as a delay in alveolar maturation at early lactation (Figure 7.1).[58] Jia *et al.*[65] demonstrated that direct exposure to BPA induced cancer cell migration through direct interaction between BPA and ITGβ1, independently of classical or non-classical estrogen receptors, highlighting the importance of the microenvironment.

7.4.2 Perinatal Exposure to BPA Alters the Transcriptional Regulation of CSN2 Expression through Epigenetic Mechanisms

The regulation of milk protein gene transcription requires both the activation of transcription factors and chromatin remodeling during the functional differentiation of the mammary gland.[55,66] In recent decades, several studies have reported that exposure to BPA induces epigenetic alterations such as DNA methylation, histone modification, and expression of non-coding RNA, altering the expression of specific genes that may persist throughout a lifetime.[5,67,68] In addition, exposure to BPA alters enzymes involved in the epigenetic regulatory machinery such as DNA methyltransferases, histone deacetylases, and histone methyltransferase in different sensitive organs.[69–72] Among them is the histone methyltransferase enhancer of zeste homolog 2 (EZH2), an estradiol-regulated gene. Exposure to BPA has increased its expression in mammary epithelial cells using both an *in vitro* and *in vivo* model.[69,71] This is significant because the main histone modification catalyzed by EZH2 is the histone H3 trimethylation at lysine 27 (H3K27me3), which is associated with gene silencing and chromatin condensation.[69] In agreement with these studies, perinatal exposure to BPA increased the mRNA expression of EZH2 during mammary secretory activation.[58]

This increased level of EZH2 expression was accompanied by an increase in H3K27me3 levels in the CSN2 promoter in BPA0.6-exposed F1 dams and in the CSN2 enhancer and exon VII in BPA0.6- and BPA52-exposed F1 dams.[58] Although DNA methylation status levels were decreased at most of the sensitive-methylation sites studied in the CSN2 promoter and enhancer, perinatal exposure to BPA induced a more condensed chromatin conformation in the CSN2 gene region by increasing EZH2 expression, and therefore H3K27me3 levels (Figure 7.1). These findings explain in part the down-regulation of CSN2 expression observed at early lactation.[58] Rijnkels *et al.*[73] have shown that the enrichment of acetyl histone 3 (H3Ac) in the regulatory region of casein genes increased their transcription rate due to a relaxed chromatin structure. In this regard, perinatal exposure to BPA decreased H3Ac levels in both the CSN2 promoter and enhancer region, which resulted in compaction of the chromatin structure and consequently, decreased the transcription of the CSN2 gene (Figure 7.1).[58] Another important fact is that the cooperation between laminin and prolactin signaling is necessary to induce histone acetylation in the CSN2 promoter.[66] Therefore, a decrease in H3Ac levels in the regulatory regions of CSN2 gene could disrupt the anchoring of laminin to the cell surface of the epithelial cells, as previously observed in BPA-exposed F1 dams during mammary secretory activation.[58] These findings indicate that changes in chromatin conformation in the regulatory regions of the CSN2 gene is an important and alternative mechanism of action of BPA on the transcriptional regulation of the CSN2 gene during secretory activation of the mammary gland.[58] Taken together, perinatal exposure to low doses of BPA induce epigenetic modifications and impair the crosstalk between laminin and lactogenic hormones signaling pathways, which seem to act synergistically to decrease CSN2 expression.

7.4.3 Direct Exposure to BPA Interferes with the Transcriptional Regulation of Milk Proteins

Previous studies have reported that exposure of F0 dams to BPA does not alter milk composition (protein, lipid, and carbohydrate content) or caloric value in rat milk samples collected at the end of lactation (LD21).[74,75] However, using an *in vitro* model, it has been shown that direct exposure to low concentrations of BPA alters the expression of both estrogen receptor alpha (ESR1) and progesterone receptor (PR) as well as the prolactin signaling pathway in mammary organoids and, therefore, modifies CSN2 and LALBA expression and its transcriptional regulation during mammary secretory differentiation (see Figure 7.2).[76] In this experiment, mammary organoids obtained and isolated from virgin female mice were cultured on Matrigel. After growing for 24 h in basal medium, the media was changed to a differentiation one (with prolactin and hydrocortisone), and the organoids were exposed to BPA for 72 h. The range of BPA concentration tested was selected by considering reported detectable levels in human biological samples and the environment.[77,78] After 72 h of induction with lactogenic hormones, mammary organoids differentiated into secretory alveoli and synthesized and secreted milk proteins into the alveolar lumen, resembling the secretory differentiation of the mammary gland at early lactation.[76] During secretory activation of the mammary gland, changes in the expression of ESR1 and PR are required for secretory activation and milk production.[2] Direct exposure to low concentrations of BPA increased both mRNA expression and the number of ESR1-positive nuclear cells. Elevated ESR1 increased the mRNA expression of the PR-B isoform and the expression of the PR protein in mammary organoids through an ESR1-dependent pathway (Figure 7.2).[76] Although no changes in PRLR and GR mRNA expression were detected in mammary organoids, Stat5a was increased and nuclear translocation of pStat5a/b was observed only at a low concentration of BPA.[76] These results were accompanied by an increase in the expression of CSN2 mRNA and protein.[76] Progesterone signaling may increase Stat5a expression and induce translocation of pStat5a/b into the nucleus, possibly mediated by the association of PR and pStat5a/b.[62,79] Therefore, the increased level of PR-B induced by exposure to BPA could increase the expression of Stat5a mRNA and cooperate with the nuclear translocation of pStat5a/b and activation of the CSN2 gene, independently of the prolactin signaling pathway (Figure 7.2).[76] In addition, in the CSN2 promoter, exposure to BPA decreased the level of DNA methylation status at a methylation-sensitive site that is close to a Stat5/GR binding site, which could also explain the increased expression of CSN2 (Figure 7.2).[76] In contrast to that which occurs with CSN2, direct exposure to low concentrations of BPA decreased the mRNA and protein expression of LALBA, and this was associated with hypermethylation at different methylation-sensitive sites in the LALBA gene (Figure 7.2).[76] It is thought that the interaction between PR and a factor (or factors) that occurs during the transduction of the prolactin signaling could be involved in the decrease of the expression of the LALBA gene.[80] Thus, the increased level of PR could be restricting the expression of LALBA in the mammary organoids.[76] Taken together, direct exposure to low concentrations of BPA affected the mammary gland's development and function in F0 dams at early lactation, altering the transcriptional regulation and expression of milk proteins. Early nutritional changes in F1 offspring, associated with alterations in F0 dams, could leave epigenetic marks impacting their mammary secretory activation.

7.5 Conclusion

Perinatal exposure to low doses of BPA modifies milk protein composition and the FA synthesis and milk FA profile, impairing milk quality and compromising the normal growth of the F2 offspring. Furthermore, the action of BPA on the expression of CSN2 during the secretory activation of the mammary gland occurred through different molecular mechanisms that act synergistically. Among them, epigenetic modifications could be a potential mechanism of BPA action in this experimental *in vivo* model that may have long-term consequences on milk quality and functional mammary differentiation in the offspring. Interestingly, using an *in vitro* model, direct

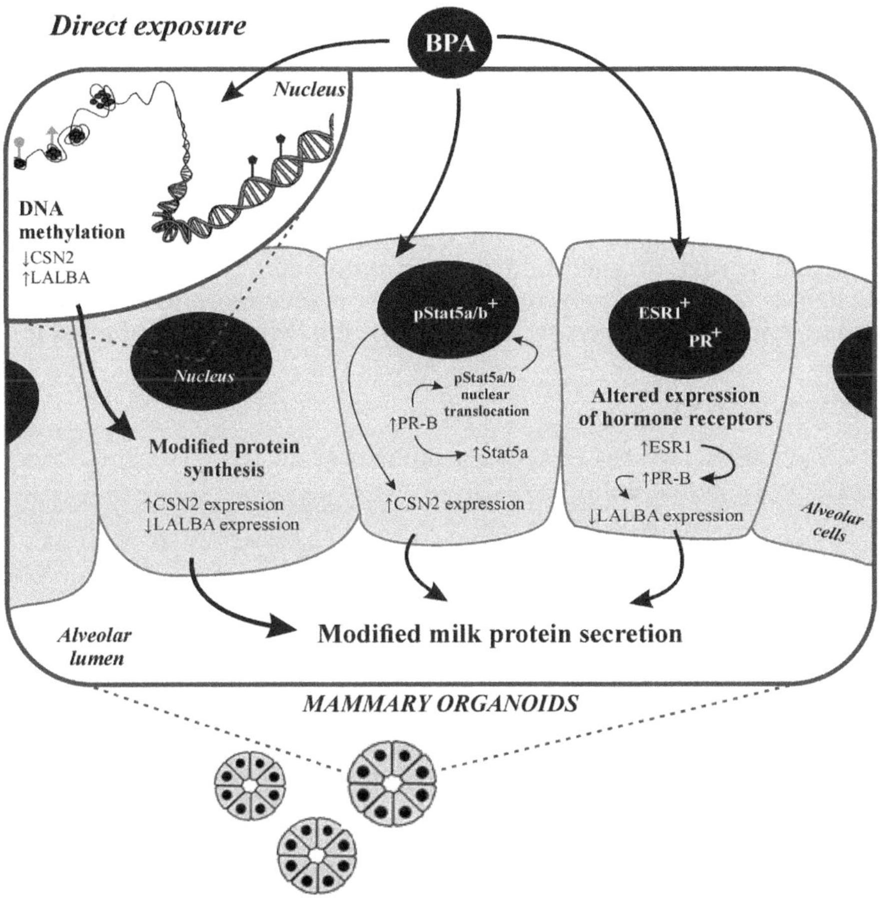

Figure 7.2 Molecular mechanism of bisphenol A (BPA) action in the functional mammary differentiation using an *in vitro* model. Direct exposure to low concentrations of BPA increased estrogen receptor alpha (ESR1) expression, which increased PR-B mRNA expression through an ESR1-dependent signaling pathway. Although PRL and GR signaling were not modified, both CSN2 and alpha-lactalbumin (LALBA) expression were altered in mammary organoids exposed to BPA. The increased level of CSN2 expression could be due to the increased PR-B expression that stimulates Stat5a mRNA expression and cooperates with the nuclear translocation of pStat5a/b and activation of the CSN2 gene. Also, high levels of CSN2 expression could be associated with hypomethylation of DNA at the CSN2 promoter. In contrast, the decrease of LALBA expression could be related to an increase in the DNA methylation status at different methylation-sensitive sites in the LALBA gene. Furthermore, increased levels of progesterone receptor expression could also restrict LALBA expression. CSN2: beta-casein; PR-B: progesterone receptor B isoform; pSTAT5 a/b: phosphorylated signal transducer and activator of transcription 5 a/b.

exposure to BPA also disrupts the mammary gland's differentiation and function during its secretory activation, including different changes in milk protein expression caused by alterations in hormonal signaling pathways or/and methylation status of their promoters. These changes are significant because they demonstrate that environmentally relevant perinatal exposure to BPA caused damage in the mammary gland. These data support the global concern about the increasing exposure to EDCs, such as BPA. They also demonstrate the high sensitivity of the offspring during the fetal and early postnatal period to any endocrine disruptor and the long-lasting consequences caused by the action of BPA on mammary functional differentiation, and consequently, on the synthesis and composition of the milk.

Abbreviations

ACCα	acetyl CoA carboxylase alpha
ADPH	adipophilin
BPA	bisphenol A
BTN	butyrophilin
CLD	cytoplasmic lipid droplet
CSN2	beta-casein
DAG1	dystroglycan 1
ECM	extracellular matrix
EDCs	endocrine-disrupting chemicals
ESR1	estrogen receptor alpha
EZH2	enhancer of zeste homolog 2
F0	dams directly exposed to bisphenol A
F1	F0 offspring
F2	F1 offspring
FA	fatty acid
FAS	fatty acid synthase
GD	gestation day
GR	glucocorticoids receptor
H3Ac	acetyl histone 3
H3K27me3	trimethyl-histone H3 (Lys 27)
ITGβ1	integrin beta 1
LD	lactation day
LPL	lipoprotein lipase
MFG	milk fat globule
PND	postnatal day
PPARγ	peroxisome proliferator-activated receptor gamma
PR	progesterone receptor
PR-B	progesterone receptor B isoform
PRLR	prolactin receptor
pSTAT5a/b	phosphorylated signal transducer and activator of transcription 5 a/b
SCD2	stearoyl-CoA desaturase

SREBP1	sterol regulatory element-binding protein 1
Stat5	signal transducer and activator of transcription 5
WAP	whey acidic protein
XOR	xanthine oxide reductase

Acknowledgements

A.L.G. is a fellow, and G.A.A and L.K. are Career Investigators of the Consejo Nacional de Investigaciones Científicas y Técnicas (CONICET), Argentina. This work was supported by a grant from the Agencia Nacional de Promoción de la Investigación, el Desarrollo Tecnológico y la Innovación (Agencia I + D + i) PICT2017 N°1532, Argentina.

References

1. R. C. Hovey, in *Encyclopedia of Reproduction*, ed. M. K. Skinner, Academic Press, 2nd edn., 2018, vol. 2, pp. 793–797.
2. S. M. Anderson, P. S. MacLean, J. F. McManaman and M. C. Neville, in *Knobil and Neill's Physiology of reproduction*, ed. T. M. Plant and A. J. Zeleznik, 4th edn., 2015, vol. 2, ch. 46, pp. 2055–2105.
3. R. J. Collier, in *Encyclopedia of Reproduction*, 1999, vol. 2, pp. 973–979.
4. C. Brisken and R. D. Rajaram, Alveolar and lactogenic differentiation, *J. Mammary Gland Biol. Neoplasia*, 2006, **11**, 239–248.
5. Y. Ma, H. Liu, J. Wu, L. Yuan, Y. Wang, X. Du, R. Wang, P. W. Marwa, P. Petlulu, X. Chen and H. Zhang, The adverse health effects of bisphenol A and related toxicity mechanisms, *Environ. Res.*, 2019, **176**, 108575.
6. B. S. Rubin, Bisphenol A: an endocrine disruptor with widespread exposure and multiple effects, *J. Steroid Biochem. Mol. Biol.*, 2011, **127**, 27–34.
7. M. Durando, L. Kass, J. Piva, C. Sonnenschein, A. M. Soto, E. H. Luque and M. Muñoz-de-Toro, Prenatal bisphenol A exposure induces preneoplastic lesions in the mammary gland in Wistar rats, *Environ. Health Perspect.*, 2007, **115**, 80–86.
8. M. Muñoz-de-Toro, C. M. Markey, P. R. Wadia, E. H. Luque, B. S. Rubin, C. Sonnenschein and A. M. Soto, Perinatal exposure to bisphenol-A alters peripubertal mammary gland development in mice, *Endocrinology*, 2005, **146**, 4138–4147.
9. A. L. Gomez, M. B. Delconte, G. A. Altamirano, L. Vigezzi, V. L. Bosquiazzo, L. F. Barbisan, J. G. Ramos, E. H. Luque, M. Muñoz-de-Toro and L. Kass, Perinatal Exposure to Bisphenol A or Diethylstilbestrol Increases the Susceptibility to Develop Mammary Gland Lesions After Estrogen Replacement Therapy in Middle-Aged Rats, *Horm. Cancer*, 2017, **8**, 78–89.
10. A. Ayyanan, O. Laribi, S. Schuepbach-Mallepell, C. Schrick, M. Gutierrez, T. Tanos, G. Lefebvre, J. Rougemont, O. Yalcin-Ozuysal and C. Brisken, Perinatal exposure to bisphenol a increases adult mammary gland progesterone response and cell number, *Mol. Endocrinol.*, 2011, **25**, 1915–1923.

11. E. C. R. Leonel, S. G. P. Campos, C. M. B. Bedolo, L. H. A. Guerra, P. S. L. Vilamaior, M. F. Calmon, P. Rahal, C. A. Amorim and S. R. Taboga, Perinatal exposure to bisphenol A impacts in the mammary gland morphology of adult Mongolian gerbils, *Exp. Mol. Pathol.*, 2020, **113**, 104374.
12. Y. Tateoka, Bisphenol A Concentration in Breast Milk following Consumption of a Canned Coffee Drink, *J. Hum. Lactation*, 2015, **31**, 474–478.
13. M. Chemek and J. Nevoral, The dark side of the breastfeeding: in the light of endocrine disruptors, *Med, J. Cell Biol.*, 2019, 7, 32–38.
14. A. Dobolyi, Novel potential regulators of maternal adaptations during lactation: tuberoinfundibular peptide 39 and amylin, *J. Neuroendocrinol.*, 2011, **23**, 1002–1008.
15. P. Palanza and L. Gioiosa, F. S. vom Saal and S. Parmigiani, Effects of developmental exposure to bisphenol A on brain and behavior in mice, *Environ. Res.*, 2008, **108**, 150–157.
16. K. G. Dewey, Maternal and fetal stress are associated with impaired lactogenesis in humans, *J. Nutr.*, 2001, **131**, 3012S–3015S.
17. G. A. Altamirano, M. Muñoz-de-Toro, E. H. Luque, A. L. Gomez, M. B. Delconte and L. Kass, Milk lipid composition is modified by perinatal exposure to bisphenol A, *Mol. Cell. Endocrinol.*, 2015, **411**, 258–267.
18. L. Kass, G. A. Altamirano, V. L. Bosquiazzo, E. H. Luque and M. Muñoz-de-Toro, Perinatal exposure to xenoestrogens impairs mammary gland differentiation and modifies milk composition in Wistar rats, *Reprod. Toxicol.*, 2012, **33**, 390–400.
19. S. Boudalia, R. Berges, C. Chabanet, M. Folia, L. Decocq, B. Pasquis, L. Abdennebi-Najar and M. C. Canivenc-Lavier, A multi-generational study on low-dose BPA exposure in Wistar rats: effects on maternal behavior, flavor intake and development, *Neurotoxicol. Teratol.*, 2014, **41**, 16–26.
20. P. L. Palanza, K. L. Howdeshell and S. Parmigiani, and F. S. vom Saal, Exposure to a low dose of bisphenol A during fetal life or in adulthood alters maternal behavior in mice, *Environ. Health Perspect.*, 2002, **110**(Suppl 3), 415–422.
21. D. Della Seta, I. Minder, F. Dessi-Fulgheri and F. Farabollini, Bisphenol-A exposure during pregnancy and lactation affects maternal behavior in rats, *Brain Res. Bull.*, 2005, **65**, 255–260.
22. M. Kundakovic, K. Gudsnuk, B. Franks, J. Madrid, R. L. Miller, F. P. Perera and F. A. Champagne, Sex-specific epigenetic disruption and behavioral changes following low-dose in utero bisphenol A exposure, *Proc. Natl. Acad. Sci. U. S. A.*, 2013, **110**, 9956–9961.
23. D. López-Rodríguez, C. F. Aylwin, V. Delli, E. Sevrin, M. Campanile, M. Martin, D. Franssen, A. Gérard, S. Blacher, E. Tirelli, A. Noël, A. Lomniczi and A.-S. Parent, Multi- and transgenerational outcomes of an exposure to a mixture of endocrine-disrupting chemicals (EDCs) on puberty and maternal behavior in the female rat, *Environ. Health Perspect.*, 2021, **129**, 87003.

24. E. Kitraki, I. Nalvarte, A. Alavian-Ghavanini and J. Rüegg, Effects of pre- and post-natal exposure to bisphenol A on the stress system, *Endocr. Disruptors*, 2016, **4**, 11.
25. S. Brummelte and L. A. Galea, Depression during pregnancy and postpartum: contribution of stress and ovarian hormones, *Prog. Neuro-Psychopharmacol. Biol. Psychiatry*, 2010, **34**, 766–776.
26. R. A. Caparros-Gonzalez, B. Romero-Gonzalez, R. Gonzalez-Perez, S. Lara-Cinisomo, P. L. Martin-Tortosa, A. Oliver-Roig and M. I. Peralta-Ramirez, Maternal and Neonatal Hair Cortisol Levels and Psychological Stress Are Associated With Onset of Secretory Activation of Human Milk Production, *Adv. Neonat. Care*, 2019, **19**, E11–E20.
27. R. Criswell, K. A. Crawford, H. Bucinca and M. E. Romano, Endocrine-disrupting chemicals and breastfeeding duration: a review, *Curr. Opin. Endocrinol., Diabetes Obes.*, 2020, **27**, 388–395.
28. N. Kasper, K. E. Peterson, Z. Zhang, K. K. Ferguson, B. N. Sanchez, A. Cantoral, J. D. Meeker, M. M. Tellez-Rojo, C. M. Pawlowski and A. S. Ettinger, Association of Bisphenol A Exposure with Breastfeeding and Perceived Insufficient Milk Supply in Mexican Women, *Matern. Child Health J.*, 2016, **20**, 1713–1719.
29. R. Chamorro-Garcia, M. Sahu, R. J. Abbey, J. Laude, N. Pham and B. Blumberg, Transgenerational inheritance of increased fat depot size, stem cell reprogramming, and hepatic steatosis elicited by prenatal exposure to the obesogen tributyltin in mice, *Environ. Health Perspect.*, 2013, **121**, 359–366.
30. H. Y. Zhang, W. Y. Xue, Y. Y. Li, Y. Ma, Y. S. Zhu, W. Q. Huo, B. Xu, W. Xia and S. Q. Xu, Perinatal exposure to 4-nonylphenol affects adipogenesis in first and second generation rats offspring, *Toxicol. Lett.*, 2014, **225**, 325–332.
31. P. Guilloteau, R. Zabielski, H. M. Hammon and C. C. Metges, Adverse effects of nutritional programming during prenatal and early postnatal life, some aspects of regulation and potential prevention and treatments, *J. Physiol. Pharmacol.*, 2009, **60**(Suppl 3), 17–35.
32. S. M. Anderson, M. C. Rudolph, J. L. McManaman and M. C. Neville, Key stages in mammary gland development. Secretory activation in the mammary gland: it's not just about milk protein synthesis!, *Breast Cancer Res.*, 2007, **9**, 204.
33. H. M. Farrel, in *Encyclopedia of Reproduction*, American Press, San Diego, 1999, pp. 256–263.
34. J. L. McManaman, Formation of milk lipids: a molecular perspective, *Clin. Lipidol.*, 2009, **4**, 391–401.
35. J. L. McManaman, T. D. Russell, J. Schaack, D. J. Orlicky and H. Robenek, Molecular determinants of milk lipid secretion, *J. Mammary Gland Biol. Neoplasia*, 2007, **12**, 259–268.
36. J. L. McManaman, C. A. Palmer, R. M. Wright and M. C. Neville, Functional regulation of xanthine oxidoreductase expression and localization

in the mouse mammary gland: evidence of a role in lipid secretion, *J. Physiol.*, 2002, **545**, 567–579.
37. T. D. Russell, C. A. Palmer, D. J. Orlicky, A. Fischer, M. C. Rudolph, M. C. Neville and J. L. McManaman, Cytoplasmic lipid droplet accumulation in developing mammary epithelial cells: roles of adipophilin and lipid metabolism, *J. Lipid Res.*, 2007, **48**, 1463–1475.
38. L. Han, M. Zhang, Z. Xing, D. N. Coleman, Y. Liang, J. J. Loor and G. Yang, Knockout of butyrophilin subfamily 1 member A1 (BTN1A1) alters lipid droplet formation and phospholipid composition in bovine mammary epithelial cells, *J. Anim. Sci. Biotechnol.*, 2020, **11**, 72.
39. N. Argov-Argaman, C. Raz and Z. Roth, Progesterone Regulation of Milk Fat Globule Size Is VLDL Dependent, *Front. Endocrinol.*, 2020, **11**, 596.
40. L. L. Listenberger, A. G. Ostermeyer-Fay, E. B. Goldberg, W. J. Brown and D. A. Brown, Adipocyte differentiation-related protein reduces the lipid droplet association of adipose triglyceride lipase and slows triacylglycerol turnover, *J. Lipid Res.*, 2007, **48**, 2751–2761.
41. J. Monks, M. S. Ladinsky and J. L. McManaman, Organellar Contacts of Milk Lipid Droplets, *Contact*, 2020, **3**, 1–9.
42. T. D. Russell, J. Schaack, D. J. Orlicky, C. Palmer, B. H. Chang, L. Chan and J. L. McManaman, Adipophilin regulates maturation of cytoplasmic lipid droplets and alveolae in differentiating mammary glands, *J. Cell Sci.*, 2011, **124**, 3247–3253.
43. T. Pavlova, Z. Spacil, V. Vidova, F. Zlamal, E. Cechova, Z. Hodicka and J. Bienertova-Vasku, Adipophilin and perilipin 3 positively correlate with total lipid content in human breast milk, *Sci. Rep.*, 2020, **10**, 360.
44. H. B. Shi, K. Yu, J. Luo, J. Li, H. B. Tian, J. J. Zhu, Y. T. Sun, D. W. Yao, H. F. Xu, H. P. Shi and J. J. Loor, Adipocyte differentiation-related protein promotes lipid accumulation in goat mammary epithelial cells, *J. Dairy Sci.*, 2015, **98**, 6954–6964.
45. L. Schipper, G. van Dijk and E. M. van der Beek, Milk lipid composition and structure; The relevance for infant brain development, *OCL*, 2020, **27**, 5.
46. J. L. McManaman and M. C. Neville, Mammary physiology and milk secretion, *Adv. Drug Delivery Rev.*, 2003, **55**, 629–641.
47. M. C. Neville and M. F. Picciano, Regulation of milk lipid secretion and composition, *Annu. Rev. Nutr.*, 1997, **17**, 159–183.
48. L. Q. Han, H. J. Li, Y. Y. Wang, H. S. Zhu, L. F. Wang, Y. J. Guo, W. F. Lu, Y. L. Wang and G. Y. Yang, mRNA abundance and expression of SLC27A, ACC, SCD, FADS, LPIN, INSIG, and PPARGC1 gene isoforms in mouse mammary glands during the lactation cycle, *Genet. Mol. Res.*, 2010, **9**, 1250–1257.
49. N. Argov-Argaman, Symposium review: Milk fat globule size: Practical implications and metabolic regulation, *J. Dairy Sci.*, 2019, **102**, 2783–2795.
50. H. Shi, J. Luo, J. Zhu, J. Li, Y. Sun, X. Lin, L. Zhang and D. Yao, PPAR gamma Regulates Genes Involved in Triacylglycerol Synthesis and

Secretion in Mammary Gland Epithelial Cells of Dairy Goats, *PPAR Res.*, 2013, **2013**, 310948.

51. G. Liang, J. Yang, J. D. Horton, R. E. Hammer, J. L. Goldstein and M. S. Brown, Diminished hepatic response to fasting/refeeding and liver X receptor agonists in mice with selective deficiency of sterol regulatory element-binding protein-1c, *J. Biol. Chem.*, 2002, **277**, 9520–9528.
52. M. C. Rudolph, J. Monks, V. Burns, M. Phistry, R. Marians, M. R. Foote, D. E. Bauman, S. M. Anderson and M. C. Neville, Sterol regulatory element binding protein and dietary lipid regulation of fatty acid synthesis in the mammary epithelium, *Am. J. Physiol. Endocrinol. Metab.*, 2010, **299**, E918–927.
53. Y. Kang, S. Hengbo, L. Jun, Z. Wangsheng, T. Huibin and S. Huaiping, PPARG Modulated Lipid Accumulation in Dairy GMEC via Regulation of ADRP Gene, *J. Cell. Biochem.*, 2015, **116**, 192–201.
54. M. Longo, F. Zatterale, J. Naderi, C. Nigro, F. Oriente, P. Formisano, C. Miele and F. Beguinot, Low-dose Bisphenol-A Promotes Epigenetic Changes at Ppargamma Promoter in Adipose Precursor Cells, *Nutrients*, 2020, **12**, 3498.
55. M. Rijnkels, E. Kabotyanski, M. B. Montazer-Torbati, C. Hue Beauvais, Y. Vassetzky, J. M. Rosen and E. Devinoy, The epigenetic landscape of mammary gland development and functional differentiation, *J. Mammary Gland Biol. Neoplasia*, 2010, **15**, 85–100.
56. J. M. Rosen, S. L. Wyszomierski and D. Hadsell, Regulation of milk protein gene expression, *Annu. Rev. Nutr.*, 1999, **19**, 407–436.
57. X. Qian and F. Q. Zhao, Current major advances in the regulation of milk protein gene expression, *Crit. Rev. Eukaryotic. Gene Expression*, 2014, **24**, 357–378.
58. G. A. Altamirano, J. G. Ramos, A. L. Gomez, E. H. Luque, M. Muñoz-de-Toro and L. Kass, Perinatal exposure to bisphenol A modifies the transcriptional regulation of the beta-casein gene during secretory activation of the rat mammary gland, *Mol. Cell. Endocrinol.*, 2017, **439**, 407–418.
59. M. Tian, Y. Qi, X. Zhang, Z. Wu, J. Chen, F. Chen, W. Guan and S. Zhang, Regulation of the JAK2-STAT5 Pathway by Signaling Molecules in the Mammary Gland, *Front. Cell Dev. Biol*, 2020, **8**, 604896.
60. B. Groner, Transcription factor regulation in mammary epithelial cells, *Domest. Anim. Endocrinol.*, 2002, **23**, 25–32.
61. F. L. Monteiro, I. Direito and L. A. Helguero, *Tissue-Specific Cell Signaling*, 2020, 279–315.
62. O. Karayazi Atici, N. Govindrajan, I. Lopetegui-Gonzalez and C. S. Shemanko, Prolactin: A hormone with diverse functions from mammary gland development to cancer metastasis, *Semin. Cell Dev. Biol.*, 2021, **114**, 159–170.
63. M. Romagnoli, L. Bresson, A. Di-Cicco, M. Perez-Lanzon, P. Legoix, S. Baulande, P. de la Grange, A. De Arcangelis, E. Georges-Labouesse, A. Sonnenberg, M. A. Deugnier, M. A. Glukhova and M. M. Faraldo,

Laminin-binding integrins are essential for the maintenance of functional mammary secretory epithelium in lactation, *Development*, 2020, **147**, dev181552.

64. M. L. Weir, M. L. Oppizzi, M. D. Henry, A. Onishi, K. P. Campbell, M. J. Bissell and J. L. Muschler, Dystroglycan loss disrupts polarity and beta-casein induction in mammary epithelial cells by perturbing laminin anchoring, *J. Cell Sci.*, 2006, **119**, 4047–4058.
65. B. Jia, T. Shi, Z. Li, S. Shan, P. Ji and Z. Li, Toxicological effects of bisphenol A exposure-induced cancer cells migration via activating directly integrin beta1, *Chemosphere*, 2019, **220**, 783–792.
66. R. Xu, V. A. Spencer and M. J. Bissell, Extracellular matrix-regulated gene expression requires cooperation of SWI/SNF and transcription factors, *J. Biol. Chem.*, 2007, **282**, 14992–14999.
67. M. Kundakovic and F. A. Champagne, Epigenetic perspective on the developmental effects of bisphenol A, *Brain, Behav., Immun.*, 2011, **25**, 1084–1093.
68. S. Singh and S. S. Li, Epigenetic effects of environmental chemicals bisphenol A and phthalates, *Int. J. Mol. Sci.*, 2012, **13**, 10143–10153.
69. A. Bhan, I. Hussain, K. I. Ansari, S. A. Bobzean, L. I. Perrotti and S. S. Mandal, Histone methyltransferase EZH2 is transcriptionally induced by estradiol as well as estrogenic endocrine disruptors bisphenol-A and diethylstilbestrol, *J. Mol. Biol.*, 2014, **426**, 3426–3441.
70. E. Dhimolea, P. R. Wadia, T. J. Murray, M. L. Settles, J. D. Treitman, C. Sonnenschein, T. Shioda and A. M. Soto, Prenatal exposure to BPA alters the epigenome of the rat mammary gland and increases the propensity to neoplastic development, *PLoS One*, 2014, **9**, e99800.
71. L. F. Doherty, J. G. Bromer, Y. Zhou, T. S. Aldad and H. S. Taylor, In utero exposure to diethylstilbestrol (DES) or bisphenol-A (BPA) increases EZH2 expression in the mammary gland: an epigenetic mechanism linking endocrine disruptors to breast cancer, *Horm. Cancer*, 2010, **1**, 146–155.
72. M. Perrot-Applanat, M. Kolf-Clauw, C. Michel and C. Beausoleil, Alteration of mammary gland development by bisphenol a and evidence of a mode of action mediated through endocrine disruption, *Mol. Cell. Endocrinol.*, 2018, **475**, 29–53.
73. M. Rijnkels, C. Freeman-Zadrowski, J. Hernandez, V. Potluri, L. Wang, W. Li and D. G. Lemay, Epigenetic modifications unlock the milk protein gene loci during mouse mammary gland development and differentiation, *PLoS One*, 2013, **8**, e53270.
74. B. S. Silva, I. M. Bertasso, C. B. Pietrobon, B. P. Lopes, T. R. Santos, N. Peixoto-Silva, J. C. Carvalho, S. Claudio-Neto, A. C. Manhaes, S. S. Cabral, G. E. G. Kluck, G. C. Atella, E. Oliveira, E. G. Moura and P. C. Lisboa, Effects of maternal bisphenol A on behavior, sex steroid and thyroid hormones levels in the adult rat offspring, *Life Sci.*, 2019, **218**, 253–264.
75. A. P. Santos-Silva, E. G. de Moura, C. R. Pinheiro, E. Oliveira and P. C. Lisboa, Short-Term and Long-Term Effects of Bisphenol A (BPA)

Exposure During Breastfeeding on the Biochemical and Endocrine Profiles in Rats, *Horm. Metab. Res.*, 2018, **50**, 491–503.
76. G. A. Altamirano, A. L. Gomez, G. Schierano-Marotti, M. Muñoz-de-Toro, H. A. Rodriguez and L. Kass, Bisphenol A and benzophenone-3 exposure alters milk protein expression and its transcriptional regulation during functional differentiation of the mammary gland in vitro, *Environ. Res.*, 2020, **191**, 110185.
77. L. N. Vandenberg, I. Chahoud, J. J. Heindel, V. Padmanabhan, F. J. Paumgartten and G. Schoenfelder, Urinary, circulating, and tissue biomonitoring studies indicate widespread exposure to bisphenol A, *Environ. Health Perspect.*, 2010, **118**, 1055–1070.
78. J. Lee, K. Choi, J. Park, H. B. Moon, G. Choi, J. J. Lee, E. Suh, H. J. Kim, S. H. Eun, G. H. Kim, G. J. Cho, S. K. Kim, S. Kim, S. Y. Kim, S. Eom, S. Choi and Y. D. Kim, Bisphenol A distribution in serum, urine, placenta, breast milk, and umbilical cord serum in a birth panel of mother-neonate pairs, *Sci. Total Environ*, 2018, **626**, 1494–1501.
79. K. A. Leehy, T. H. Truong, L. J. Mauro and C. A. Lange, Progesterone receptors (PR) mediate STAT actions: PR and prolactin receptor signaling crosstalk in breast cancer models, *J. Steroid Biochem. Mol. Biol.*, 2018, **176**, 88–93.
80. L. Sankaran and Y. J. Topper, Progesterone and prolactin are both required for suppression of the induction of rat alpha-lactalbumin activity, *Biochem. Biophys. Res. Commun.*, 1988, **155**, 1038–1045.

Subject Index